DISCRETE EVENT PHYSICS

DISCRETE EVENT PHYSICS

Volume II

WILLIAM DELANEY

iUniverse, Inc.
Bloomington

Discrete Event Physics
Volume II

iUniverse books may be ordered through booksellers or by contacting:

iUniverse
1663 Liberty Drive
Bloomington, IN 47403
www.iuniverse.com
1-800-Authors (1-800-288-4677)

ISBN: 978-1-4620-6501-1 (sc)
ISBN: 978-1-4620-6502-8 (ebk)

Printed in the United States of America

iUniverse rev. date: 02/24/2012

CONTENTS

PREFACE

Volume II introduces new general concepts to Discrete Event Physics (in Chapters 1, 2 and 7), and presents applications of the theory in the other chapters.

Chapter 1 presents and compares partial definitions of sequential, causal and hierarchical relations.

Chapter 2 discusses frame definition in terms of various kinds of events and the classification of frames in terms of properties involved, symmetry considerations and interframe relationships.

Chapter 3 presents detailed definitions of properties traditionally associated with space time such as coordinates and intervals, velocity and acceleration—all in terms of discrete events underlying them.

Chapter 4 extends proof of the limitations of The Operational Paradigm regarding the use of measurement procedures to define properties: it demonstrates that no property can be defined using procedures.

Chapter 5 resolves the long standing problem of probability definition in science by presenting a general such definition in the Discrete Event Physics paradigm.

Chapter 6 presents a definition of waves, their properties, and their interactions in terms of their underlying events, with detailed attention to the sub structure of the waves and that of their properties. Locality considerations are also discussed.

Chapter 7 defines the concepts of objective meaning, knowledge and understanding in the Discrete Event Physics paradigm.

Chapter 8 discusses thermodynamic energy and entropy as properties of discrete event processes and process trees. Energy and entropy definitions are proposed and compared to each other and to the discrete event definition of time duration.

Chapter 9 presents a solution to the ERP Paradox.

Chapter 10 discusses locality in double slit interference experiments.

1

DISCRETE EVENT PHYSICS:

UPDATES TO BASIC THEORY

1. Introduction

Section 2 presents partial definitions of sequential, causal and hierarchical relations and Section 3 compares those relations.

2. Definitions of sequential, causal and hierarchical relations

In (Delaney, 2005) events are defined as sets whose constituents are sub events or *occurences*. There are two basic types of events:

- I/O events = events defined in terms of their inputs and corresponding outputs
- O-I events = events defined in terms of two I/O events such that the output from one of them is the input to the other one. Such events are *relations* between their constituent I/O events and are also called O-I relations, or more simply just relations.

This section defines various types of inter event O-I relations: *sequential* ones in Definition 1, *causal* ones in Definition 2, and *hierarchical* ones in Definition 3.

Definition 1. Sequential O-I relations

A sequential O-I relation <EVa, EVb> is one in which the output from EVa becomes the input to EVb, which is understood to correspond to an ordering relation in which

1. EVa *precedes* EVb and EVb *is preceded by* EVa
2. EVa and EVb are at the same (hierarchical) *level* in the sense that
   a) EVa is not a sub event of EVb and EVb is not a sub event of EVa
   b) EVa and EVb are non intersecting (do not have sub events in common)
3. EVa precedes EVb is *inconsistent with* EVb precedes EVa
4. EVa is not necessarily the *cause* of EVb

Sequential O-I relations can be concatenated to form O-I *processes*, e.g. <EVa, EVb> <EVb, EVc>, the latter being themselves relations between their first and last events (respectively EVa and EVc in the example).

Such relations are structurally similar to the mathematical models of property value sequences studied in Special Relativity, such as spatial and temporal coordinate sequences along a trajectory traversed by a moving object.

Definition 2. Causal O-I relations

A *causal* O-I relation <EVa, EVb> is one in which the output from EVa becomes the input to EVb, which it causes, or contributes to cause. It is understood to correspond to an ordering relation in which

1. EVa *causes* EVb and EVb *is caused by* EVa
2. EVa and EVb are at the same (hierarchical) *level* in the sense that
   a) EVa is not a sub event of EVb and EVb is not a sub event of EVa
   b) EVa and EVb are non intersecting (do not have sub events in common)
3. EVa causes EVb is *inconsistent with* EVb causes EVa
4. EVa *precedes* EVb and EVb *is preceded by* EVa

Causal O-I relations can be concatenated to form causal O-I processes just as was discussed above for sequential relations.

Definition 3. Hierarchical O-I relations

A *hierarchical* O-I relation <EVa, EVb> is one in which the output from EVa becomes the input to EVb, which it *contains* as a sub event. It is understood to correspond to an ordering relation in which EVa contains EVb as its sub event so that—capitalizing CONTAIN for consistency with the convention adopted in (Delaney, 2005):

1. EVa *CONTAINS* EVb and EVb *IS CONTAINED IN* EVa
2. EVa and EVb are at different (hierarchical) *levels* in the sense that
   a) EVb is a sub event of EVa
   b) EVa and EVb are maximally intersecting in the sense that EVa contains EVb in its entirety—not just some of its sub events
3. EVa CONTAINS EVb is *inconsistent with* EVb CONTAINS EVa

Hierarchical O-I relations can be concatenated to form hierarchical O-I processes just as was done for the other relations discussed above.

3. Comparisons

This section compares the types of inter event relations defined in section 2.

The above definitions 1 and 2, respectively of sequential and causal relations, are very similar, differing only in Item 4 of Definition 1. Accepting or rejecting this difference has important consequences:

• accepting the just cited Item 4, causality appears to be a special case of sequentiality, i.e. an event can be

preceded by another without the latter being one of its causes.
- rejecting Item 4, sequential relations can be understood to be a consequence of causal ones: causes are always the physical reason for event sequences

The similarity between sequential and causal relations can suggest resolutions to seemingly paradoxical interpretations of results of experiments like the ERP thought experiment (Einstein A, Podolsky B, Rosen N, 1935) and experiments for testing Bell's locality hypothesis (Bell1 J S, 1966) in which events EVa and EVb occurring at widely separated spatial locations seem to be related even though their temporal separation would not permit a causal relation consistent with slower-than-light communication between them. Instead of a causal relation a sequential relation can be hypothesized, as is often done in relativity theory in the interpretation of relations (to events called reference frames) of events 'having' spatial coordinates, velocity, momentum, etc.

Causal and hierarchical relations as defined respectively in Definitions 2. and 3, can be compared as follows:

- causal relations <EVa,EVb> between events EVa and EVb can be hypothesized to contribute to the formation of hierarchical relations <EVa,EVc> and <EVb,EVc> where EVc is a SUPER event that CONTAINS EVa and EVb—such hierarchical relations are said to be *bottom up* ones because they are directed from EVa and EVb to their SUPER event EVc
- conversely, hierarchical relations <EVa,EVc> and <EVb,EVc> can be hypothesized to contribute to the formation of causal relations <EVa,EVb> between events EVa and EVb—such hierarchical relations are said to be *top down* ones because they are directed from EVc to its SUB events EVa and EVb. Top down hierarchical relations can be used as a basis for *inheritance* phenomena whereby SUB events obtain

property values from their SUPER event, for example initial coordinates for a trajectory which establish the collocation of its beginning 'in space-time' (Delaney, 2005, Chapters 4 and 5). In the limit, a hierarchical process consisting of consecutive such top down relations could extend from a space-time origin of the universe to a specific event, e.g. EVa or EVb, so as to collocate it in that universe.

2

REFERENCE FRAMES IN DISCRETE EVENT PHYSICS

1. Introduction

A fundamental concept in relativity theory is that of frame of reference. Different such frames can be distinguished as having different points of origin or axis directions, or by involving different relative motions. Special Relativity concentrates on inertial frames (an idealization, since all physical systems are accelerated to some degree, at least by gravity), whereas General Relativity can also accommodate non inertial (accelerated) frames.

In the context of the study of a specific system or event, a frame is typically chosen by an investigator in such a way to simplify the study, not because the frame is inherently "better" than another—there are no "privileged" frames.

The nature of properties and their values can be characterized in different ways: by the property being called a relative property (e.g. a relative coordinate), by saying that its value is relative to a frame or existent, or by saying that a property is a property of a relation (between events). The latter perspective facilitates the definition of frame hierarchies in a particularly simple way.

Frames can be specified at different levels of detail. For example, a frame specification may be definite (expressed in terms of specific existents) or indefinite (expressed just in terms of existent types). Also, for certain purposes it is sufficient to simply specify the 'type' of a frame, e.g. that it is inertial, that it has certain symmetry properties, that it is a frame for certain properties, or that it is related to certain other frames in specific ways (see Section 3.2 for a more general discussion of frame types).

Section 3.1 discusses frame definition in terms of events, event pairs (relations) and general event sequences (PART's). Section 3.2 illustrates the classification of frames according to a combination of three criteria: properties involved, symmetry considerations and interframe relationships.

3.1 Frame definition

This section presents frame definition in terms of single events and event pairs in Sub Section 3.1.1, and in terms of general event sequences in Sub Section 3.1.2.

3.1.1 Frame definition in terms of events and event pairs

The following table defines forms used in Discrete Event Physics to define property values of events and relations, taking frame dependence into account.

Table I. Definite and indefinite forms for events and anti symmetric relations having frame dependent property values. Items with suffix 'a' and 'b' specify the frame dependence in terms of a relation between events and those with suffix 'c' specify it using the symbol FR= in the value specification (after '.V') with

a) FR=EVe := the EV 'e' in the role of a frame of reference

b) FR=STEVe := the space-time event (STEV) 'e' in the role of a frame of reference

c) <EVy,EVx>.Pp.V := relation <EVy,EVx> as a whole, having the property p with value v

d) <FR=EVy,EVx>.Pp.V := EVx has the property p with value v (in frame EVy)

e) <a,b> := an antisymmetric relation between a and b

f) (a,b) := a symmetric relation between a and b

g) ⟨a,b⟩ := a relation between a and b with unspecified symmetry properties

then

	Forms	Meaning
1	EVx.Pp.V	event x having property p with some value (in some unspecified frame—if the property requires a frame)
2	EVx.Pp.Vv	event x having property p with value 'v' (in some frame)
3	{EV.Pp.Vv}	the set of all [relations or] EV's with value v of property p, irrespective of frame considerations—the definition set of value v
4a	⟨EVy,EVx⟩.Pp.Vv	a relation existing between EVy and EVx and having property p with value 'v'. If a frame is necessary the relation as a whole is the frame.

4b	⟨FR=,EVx⟩.Pp.Vv	a relation involving event 'x' and having property p with value 'v' in an unspecified frame of unspecified type
4c	EVx.Pp.V(FR=,v) or EVx.Pp.Vv	event x having property p with value 'v', in a frame of unspecified type (if one is necessary)
5a	<STEV,EVx>.Pp.Vv	a relation existing between EVx and some STEV and having property p with value v
5b	<FR=STEV,EVx>.Pp.V	a relation existing between EVx and some STEV and having property p with value v in the frame of that STEV
5c	EVx.Pp.V(FR=STEV,V)	event x having property p with value v in a frame of type STEV
6a	{<STEV,EV>.Pp.Vv}	the set of all relations: 1) existing between events and STEV's and 2)having property p with value v
6b	{<FR=STEV,EV>.Pp.Vv}	the set of all relations existing between events and STEV's and having property p with value v in the frame of the STEV
6c	{EV.Pp.V(FR=STEV)}	the set of all events having property p with some unspecified value in a frame of type STEV

7a	<STEV,EVx>.Pp.Vv	a relation existing between a STEV and the event 'x' and having property p with value 'v'
7b	<FR=STEV,EVx>.Pp.Vv	a relation existing between a STEV and the event 'x' and having property p with value 'v' in the frame of that STEV
7c	EVx.Pp.V(FR=STEV,v)	event x having property p with value 'v' in a frame of type STEV
8a	{<STEV,EV>.Pp.Vv}	the set of all relations existing between STEV's and EV's and having property p with value v
8b	{<FR=STEV,EV>.Pp.Vv}	the set of all relations existing between STEV's and EV's and having property p with value v in the frame of the STEV
8c	{EV.Pp.V(FR=STEV,v)}	the set of all EV's with value v of property p, relative to some unspecified frame of type STEV
9a	⟨STEVe,EVx⟩.Pp.Vv	a relation existing between STEVe and event x and having property p with value 'v'
9b	⟨FR=STEVe,EVx⟩.Pp.Vv	a relation existing between STEVe and event x and having property p with value 'v' in the frame STEVe

9c	EVx.Pp.V(FR=STEVe,v)	event x having property p with value 'v' in frame STEVe
10a	{⟨STEVe,EV⟩.Pp.Vv}	the set of all relations existing between STEVe and an EV and having value v of property p
10b	{⟨FR=STEVe,EV⟩.Pp.Vv}	the set of all relations existing between STEVe and an EV and having value v of property p in the frame of STEVe
10c	{EV.Pp.V(FR=STEVe,v)}	the set of all EV's with value v of property p in frame STEVe—the subset of {EV.Pp.Vv} defining v relative to frame STEVe
11a	⟨EVz,⟨EVy,EVx⟩⟩.Pp.Vv	a relation existing between EVz and the relation ⟨EVy,EVx⟩ and having property p with value 'v'
11b	⟨FR=EVz,<EVy,EVx>⟩. Pp.Vv	a relation existing between EVz and the relation ⟨EVy,EVx⟩ and having property p with value 'v' in the frame of EVz
11c	⟨EVy,EVx⟩. Pp.V(FR=EVz,v)	relation ⟨EVy,EVx⟩ having property p with value 'v' in frame EVz

The above definitions can be converted to definitions with specific symmetry properties:

- for anti-symmetric relations, by changing the pairs ⟨STEV,EVx⟩ in column 'Forms' into ordered pairs

<STEV,EVx>, corresponding only to relations directed <u>from</u> the first element (STEV) of the relation <u>to</u> the second one (EVx)

- for symmetric relations, by changing the ⟨STEV,EVx⟩ in column 'Forms' into unordered pairs (STEV,EVx), corresponding to relations directed <u>from</u> the first element (STEV) of the relation <u>to</u> the second one (EVx) <u>and vice-versa</u>

Items 1 and 2 of the above table can be very useful for avoiding discussions (e.g. definitions) containing many expressions with references to a frame in the form of a structured event. It can be much simpler to just preface the whole discussion with a specification of a pertinent frame or frame type.

3.1.2 Frame definition in terms of general event sequences

Above, the concept of frame dependent value was formalized in terms of events (EV's) and event pairs (⟨EV,EV⟩). For enhanced precision it can be useful to specify the involved events as occurrences within a certain context. For this purpose the FRSEQ construct is useful: frames are formalized in terms of sequences of events in the context of larger event sequences, the former sequences being called FRSEQ's of the larger ones. The inter-event relation corresponding to a frame is then collocated within a certain FRSEQ such that the first event in the FRSEQ is the frame of reference for the last one, or in other words properties correspond to a relation between the first and last events. The immediately following definition of frame related concepts makes use of examples involving FRSEQ's in the context of trajectories in space-time (PxtTRJ's), using interaction pairs (STEV's) as primitive events.

Definition 1a. Contextualization of reference frames using event sequences

with

- EV := the type of the events being used as primitives in the definition of a more complex event (e.g. EV could be PA, SEV, STEV)
- EVw(k) := the k-th primitive event in a sequence of such events
- a symmetric relation between physical events := a relation that implies the physical existence of its inverse (as along trajectory sense of motion does, for example)
- STEVw(j) = ⟨EVw(j),EVw(j+1)⟩, which is a relation between EVw(j) and EVw(j+1): that is symmetric if '⟨'and'⟩' 'are replaced respectively by' ('and')' and is antisymmetric if they are replaced by '<'and'>'
- FRSEQw(j,k) := the sequence of O-I relations <EVw(j), EVw(j+1)>, . . . , <EVw(k-1), EVw(k)>; such a sequence may correspond to the motion of some system through the EV's from EVw(j) to EVw(k). When EV is a STEV, a FRSEQ consists of the STEV's contained, for example, in a PxtTRJ and is then itself a PxtTRJ; thus PxtTRJu(m):FRSEQw(j,k) is the FRSEQ of PxtTRJu beginning at the jth STEV of PxtTRJu and ending at its kth STEV. Such a FRSEQ is a complex O-I relation between its beginning and end STEV's
- relational property := a property of a relation (such as the relation FRSEQw(j,k))

then

1.   For general reference frames
FRSEQw(j,k):Pp:V

:=

the value of property 'p' of the relation FRSEQw(j,k)
which is similar to
the value of property 'p' of EVw(k) relative to (in the frame of) EVw(j)

2. Explicit reference frames in the context of a complex event PxtTRJu

$$PxtTRJu(n):FRSEQw(j,k):Pp:V$$

$$:=$$

the value of property 'p' of the relation FRSEQw(j,k) between STEVw(k) and the event STEVw(j) of PxtTRJu(n)

3. Explicit reference frames at the beginning of a complex event PxtTRJu

$$PxtTRJu(n):FRSEQw(0,k):Pp:V$$

$$:=$$

the value of property 'p' of the relation FRSEQw(0,k) between STEVw(k) and the beginning event STEVw(0) of PxtTRJu(n)

which is similar to

the value of property 'p' of the relation FRSEQw(0,k) between STEVw(k) and the beginning event STEVw(0) of PxtTRJu(n)

in the reference frame of said beginning event

4. Implicit reference frame at the beginning of a complex event PxtTRJu

$$PxtTRJu(n):STEVw(k):Pp:V$$

is synonymous with

$$PxtTRJu(n):FRSEQw(0,k):Pp:V$$

Those familiar with Discrete Event Physics will recognize that the above definition defines the FRSEQ construct exactly the way the PART construct is defined in (Delaney, 2004, 2005), i.e. as a generalization of STEV's to longer event sequences like RTEV's (consecutive STEV pairs), etc. The new symbol FRSEQ was introduced here especially to avoid confusion that could arise when considering frames for properties of PART's. Examples of such frames are shown in the following definition.

Definition 1b. Frames for properties of PART's

with

- the contents of Definition 1a
- PART's defined exactly like FRSEQ's in Definition 1a

then

1. Explicit reference frames involving PART's of the same PxtTRJ

$$<PARTw(i,j), PARTu(l,k)> :Pp:V$$

$$:=$$

the value of property 'p' of the relation
$$<PARTw(i,j),PARTu(l,k)>$$
relating PARTu(l,k) to PARTw(i,j)

2. Explicit reference frames involving PART's of different complex events

$$<FR=PxtTRJw(m):PARTw(i,j), PxtTRJu(n):PARTu(l,k)>$$
$$:Pp:V$$

$$:=$$

the value of property 'p' of the relation <PxtTRJw(m):
PARTw(i,j),PxtTRJu(n):PARTu(l,k)>
said relation existing between PxtTRJu(n):PARTu(l,k)
and PxtTRJw(m):PARTw(i,j) with the former being
the value's reference frame

3. Explicit reference frames involving PART's of different complex events
in the context of their common SUPER-PxtTRJ,
PxtTRJq(i)

$$PxtTRJq:<PxtTRJw(m):PARTw(i,j),PxtTRJu(n):PARTu(l,k)> :Pp:V$$

$$:=$$

the value of property 'p' of the relation <PxtTRJw(m):
PARTw(i,j),PxtTRJu(n):PARTu(l,k)>
relating PxtTRJu(n):PARTu(l,k) to
PxtTRJw(m):PARTw(i,j) with PxtTRJq in the context
of PxtTRJq

3.2. Frame type classification

Here details are presented concerning the formal definition
of frame types as introduced in Section 2, first with regard,
in subsection 3.2.1, to considerations of symmetry and
involved property, and then with regard, in subsection 3.2.2,
to interframe relationships.

3.2.1 Frame type classification: symmetry and involved property

The following table illustrates frame type classification
with respect to a combination of criteria: symmetry (of the
relationship between event pairs) and property involved. Anti
symmetrical relationships have properties generically called
coordinates and symmetrical ones have properties called
intervals.

Table II. Frame type classification: general forms and
examples for specific properties. The text between square
brackets '[]' defines the reference frame for a property value
in terms of one of the events appearing in the relation under
consideration—ignoring said text implies that the reference
frame (if any) is to be understood to correspond to the whole
relation under consideration.

with		
	EV as a generic symbol for events of any complexity—not just simple events	
then		

General Forms		
Value	Definition set for value	Frame type
val	{<EVa, EV> .Pcoordinate. Vval} := all binary relations between events with property 'coordinate' having value 'val' [relative to (in the frame of) fixed EV 'a']	val-coordinate
nil	{<EVa, EV> .Pcoordinate. Vnil} := all relations between events with property 'coordinate' having value 'nil' [relative to (in the frame of) fixed EV 'a']	nil-val-coordinate
val	{(EV, EV) .Pinterval.Vval} := all binary relations between events with interval (distance) value 'val' relative to each other	val-interval
nil	{(EV, EV) .Pinterval.Vnil} := all binary relations between events with interval (distance) value 'nil' relative to each other	nil-val-interval
Examples		

with EV as a generic symbol for events of any complexity—not just simple events **then**		
x	{<EVa, EV> .Pposition-magnitude.Vx} := all binary relations between events with x-position-magnitude value x [relative to (in the frame of) fixed EV 'a']	x-position-magnitude
x	{<EVa, EV> .Pposition-magnitude.Vnil} := all binary relations between events with x-position-magnitude value 'nil' [relative to (in the frame of) fixed EV 'a']	nil-x-position-magnitude (co-incident EV's)
x	{(EV, EV) .Pdistance.Vdx} := all binary relations between events with distance value dx	distance
x	{(EV, EV) .Pdistance.Vnil} := all binary relations between events with distance value 'nil'	nil-distance (congruent/co-spatial EV's)
t	{<EVa, EV> .Ptime.Vt} := all EV's with time value t [relative to (in the frame of) fixed EV 'a']	t-coordinate

with		
	EV as a generic symbol for events of any complexity—not just simple events	
then		
t	{<EVa, EV> .Ptime.Vnil} := all EV's with time value nil [relative to (in the frame of) fixed EV 'a']	nil-t-coordinate (simultaneous EV's)
t	{(EV, EV) .Ptime interval. Vt} := all EV pairs with time interval value t relative to each other	t-interval
t	{(EV, EV) .Ptime interval. Vnil} := all EV pairs with time interval value nil relative to each other	nil-t-interval (co-existent EV's)
v	{<EVa,EV> .Pvelocity.Vv} := all EV's with velocity value v [relative to (in the frame of) fixed EV 'a']	v-coordinate
v	{<EVa,EV> .Pvelocity.Vnil} := all EV's with velocity value nil [relative to (in the frame of) fixed EV 'a']	nil-v-coordinate co-moving with EVa

with	EV as a generic symbol for events of any complexity—not just simple events	
then		
v-mag	{(EV,EV) .Pvel-mag. Vv-mag} := all EV pairs with velocity magnitude value v-mag [relative to (in the frame of) each other]	v-mag-interval
v-mag	{(EV,EV) .Pvel-mag.Vnil} := all EV pairs with the velocity magnitude value nil [relative to (in the frame of) each other]	nil-v-mag-interval

3.2.2 Frame type classification: interframe relationships

As mentioned above, frame types can also be characterized in terms of interframe relationships and their affect on property values. A simple example is that of frames that differ only by having diverse spatial orientations:

- any two such frames are related by angles (θ,φ) whose values are those required to rotate one of the frames into coincidence with the other one
- spatial intervals and momentum magnitudes will remain invariant under frame change
- the values of the spatial coordinates and momentum components associated with a system will generally vary with the frame from which they are observed
- the values of spatial coordinates in one frame can be transformed into those in another frame by applying a rotation matrix (whose elements are functions of

the angles (θ, φ)) to a vector whose elements are the coordinates—the same matrix being also applicable to a momentum vector

The following table lists the above and other examples of frame relations in a form designed to facilitate comparisons.

Table III. Examples of relations between frames. Each row corresponds to a specific example: column 1 gives its name, column 2 characterizes the relationship between different frames (sometimes in terms of parameters), column 3 lists invariants under frame change, column 4 lists some property values that may vary with choice of frame, column 5 characterizes the relationship between values in different frames in terms of a transformation function.

Frame relations / Example	Interframe relationship and its parameterization	Interframe invariants	Variable property values	Property value interframe transformation
1. 3d rotation	parameter $\theta :=$ interframe angle	x magnitude p magnitude	coordinates x_i, $i\in[1,3]$, — momentum components p_i, $i\in[1,3]$	$(x')=f(x',\theta)$ $(p')=f(p',\theta)$ f implements rotation matrix
2. 4d special relativity	$v :=$ interframe velocity magnitude	—space time interval —mass —action	x_i, $i\in[1,4]$ p_i, $i\in[1,4]$	$(x')=f(x',v)$ $(p')=f(p',v)$ f implements Lorentz transformation

3. 4d general relativity	(v,a,,):= inter-frame (velocity, acceleration) magnitudes, .	−local space−time interval −mass ? −action ?	x_i, i∈[1,4] p_i, i∈[1,4] also 4,4 tensors x_{ij}, p_{ij}	(x')=f(x,v,a) (p')=f(p,v,a) f implements general coordinate, momentum transformation
4a. DEP (Discrete Event Physics)	event pair, e.g. <EV3,EV2>	events, and any properties whose values do not change under frame change for all frame pairs <EVi,EVj> ∀i,j	property values	<EV3,EV1>:Pp:Vv' = f(<EV2,EV1>:Pp:Vv, <EV3,EV2>) v,v' are values of property p of relations <EV2,EV1> and <EV3,EV1> and <EV2,EV3> = the relation between EV2 and EV3>
4b. DEP	property value of an event pair, e.g. <EV3,EV2>:Pb:Vq := value q of property b of the relation <EV3,EV2> between events EV3 and EV2	events, and any properties whose values do not change under frame change, for all frame pairs <EVi,EVj> ∀i,j	property values	<EV3,EV1>:Pp:Vv' = f(<EV2,EV1>:Pp:Vv,q) v,v' being values of property p of relations <EV2,EV1> and <EV3,EV1> and q is as defined in column 2.

* The reference to Mathematical Quantum Mechanics in rows 4a,b of the above table is meant to suggest that the Discrete Event Physics approach to defining frames in terms of events might help to solve problems arising in that discipline because of difficulties in reconciling the uncertainty principle with the definition of frames in terms of properties.

Because they refer to values that are not necessarily numerical, Items 4a,b refer to a more general level of precision than Items 1,2,3—aside from this they seem to include the latter as special cases.

3

SPACE TIME: UNDERLYING EVENTS AND THEIR PROPERTIES

1. Introduction
This chapter presents detailed definitions of properties traditionally associated with space time such as coordinates and intervals, velocity and acceleration—all in terms of discrete events underlying them.

2. General theory
The organization of the presentation is based on the step wise refinement methodology typical of Discrete Event Physics: for each property a very general definition is enunciated and then details are added one step at a time. The initial steps, in Sections 3, 4, and 5—respectively for coordinates, velocity, and acceleration—are aimed at formalizing the properties in the language of Discrete Event Physics. Then more details are added in Section 6 on the basis of prototypical measurement procedures.

2.1 Symbolism for space time related events
In the following, events having the properties of space time extension, velocity, and acceleration will be discussed and defined. For reader convenience the symbolism to be adopted for this purpose is outlined in the following Table I.

Table I. Space time related events, their properties and property values. The "i" in i-vel and i-acc means instantaneous, "-mag", means magnitude (value ignoring sense) and the suffix "-v" means value. The concepts of trajectory sense and translation sense are defined in (Delaney, 2004, 2005).

Event type	Property	Value
PxtTRJ	xt := space-time coordinate along a trajectory = ((x-sense, x-mag),t) where x-sense=trj-sense := along trajectory sense	(x,t) = ((x-sense-v, x-mag-v),t)
PxtTRJ. STEV	xt-change:= space-time coordinate change = (x-change, t-change) where x-change = (x-change-sense, x-change-mag) where x-change-sense := translation sense	(dx,dt) = ((dx-sense, dx-mag),dt) = ((x-change-sense-v, x-change-mag-v),dt)
PxtTRJ	sti := space-time interval along a trajectory	sti-v := the value of sti
PxtTRJ. STEV	sti-change space-time interval change along a trajectory	sti-change-v := the value of sti-change

Event type	Property	Value
PxtTRJ. STEV	i-vel := instantaneous velocity = ((i-vel-sense, i-vel-mag))	iv = (iv-sense, iv-mag)) = ((i-vel-sense-v, i-vel-mag-v)) iv∈{dx,dt)}:SUB iv-mag ∈ {(dx-mag,dt)}:SUB
PxtTRJ	vel := velocity = (vel-sense, vel-mag)	v=(v-sense, v-mag) = (vel-sense-v, vel-mag-v) v ∈ {x,t}:SUB
PxtTRJ. RTEV	v-change := velocity change = (v-change-sense, v-change-mag)	(dv,dt) = ((dv-sense, dv-mag),dt) = ((v-change-sense-v, v-change-mag-v),dt) v-change-mag-v ∈ {(dx-mag, dt)}:SUB
PxtTRJ. RTEV	i-acc := instantaneous acceleration = (i-acc-sense, i-acc-mag)	ia = (ia-sense, ia-mag) = (i-acc-sense-v, i-acc-mag-v) ia ∈ {dv,dt)}:SUB ia-mag ∈ {(dv-mag,dt)}:SUB
PxtTRJ	acc:= acceleration = (a-sense, a-mag)	a = (a-sense-v, a-mag-v) a∈{v,t}:SUB

3. Space time

(Delaney, 2004, 2005, Chapter 5, Section 3.6) introduced a definition of *space time process trajectories* (PxtTRJ's) involving both endurance in time and (motion induced) extension in space. Such trajectories have the property of space time extension (or coordinate) "xt" with values (x,t); xt and its values are defined in terms of the actually existing trajectories having them as illustrated in the following Table II.

Table II. Basic definitions of space time extension

Expression	Meaning
with	
{PxtTRJ.Pxt.V(x,t)}:= the set containing all PxtTRJ's with xt value "(x,t)" (the definition set for that value)	
{PxtTRJ.Pxt.V(x,t)}:inv := the invariant associated with the definition set for value "(x,t)" of the xt property of a PxtTRJ	
then	
Definition 2a. General xt Definition	
{PxtTRJ.Pxt}:inv	the characteristic invariant of the set whose elements are PxtTRJ's having <u>some</u> xt value
{PxtTRJ. Pxt.V(x,t}:sub:inv	the invariant associated with the subset of all PxtTRJ's having value (x,t), i.e. (x,t) itself
Definition 2b. xt as a function	
F\xt(PxtTRJw)	a function associating an xt value with an PxtTRJ, e.g. F\xt(PxtTRJw) = {PxtTRJ.Pxt.V(x,t}:sub:inv

A PxtTRJ can be understood to consist of a sequence of *space time events* (STEV's) or a hierarchy of SUB-PxtTRJ's. As explained in the above cited (Delaney, 2004, 2005, Chapter 5, Section 3.6), the STEV's consist of spatial events (SEV's) involving change in position, and temporal events (TEV's) involving change in time as a PxtTRJ is traversed. The SEV's and TEV's are ordered pairs of events having event type PA. A PA is an input-ouput (I/O) event and is referred to as an *interaction* or more generally as an *activity*.

A SEV "x" corresponds to an output from an interaction PAy that is input to another interaction PAz, i.e. it has the form

$$SEVx = <PAy,PAz>, x \neq y \,.$$

When the output corresponds to a physical system (PS) the SEV corresponds to the motion from PAy to PAz. It is important to understand that

by definition the constituent PA's of a SEV are to be understood to have no actually existing PA's between them on a trajectory containing the SEV, as emphasized in (Delaney, 2004, 2005).

A SEV has the property of *spatial change*: it can be detailed in the form SEV.Px-change.Vdx where dx is the value of the spatial change associated with a motion along a space-time trajectory.

A TEV corresponds to a *self interaction* in which an output from activity PAy is input to that same activity, i.e.

$$TEVx = <PAy,PAy>$$

A TEV has the property of *time change*: it can be detailed in the form TEV.Pt-change.Vdt where dt is the value of the time change associated with a motion along a space-time trajectory.

An important difference between SEV's and TEV's is that x-change is oriented, but t-change is not, so that (see Table I above):

$$x\text{-change}=(x\text{-change-sense, }x\text{-change-mag})$$

Thus the x-change value dx = x-change-v is

$$dx= <dx\text{-sense, }dx\text{-mag}>$$

In mathematical models, where dx is a number, dx-sense is typically represented by prefixing a + or - sign to the dx-mag value.

Generalizing the approach in ((Delaney, 2004, 2005, Chapter 5, Section 3.6), the SEV's and TEV's can be combined into space-time events (STEV's) as follows:

$$STEVw(n) \in \{SEVe(n), TEVd(n), <SEVe(n)|TEVd(n)>, <TEVd(n)|SEVe(n)>\}$$ so that a STEV may be

1. a SEV having the x-change property with value dx: SEV.Px-change.Vdx
2. a TEV having the t-change property with value dt: TEV. Pt-change.Vdt
3. a SEV followed by a TEV with an x-change and a t-change occurring in that order
4. a TEV followed by a SEV with a t-change and an x-change occurring in that order

Taking into consideration all the just stated possibilities, a STEV as a whole has the property xt-change with values (dx,dt) defined as:

$$(dx,dt) = <dx,dt> \text{ .eor. } <dt,dx>$$

where either dx or dt (but not both) can be nil, and .eor. is the "exclusive or" operation.

In terms of STEV's a PxtTRJ is defined as follows:

Definition 3a. PxtTRJ's, STEV's

with

- SEVa(n)==<PAb(n),PAc(n)>, b≠c and PAc(n) is the immediate-successor of PAb(n) in PxtTRJw(n)
- TEVa(n)==<PAb(n),PAb(n)>
- nil:= nothing, empty, zero, . . .
- STEVw(n) ∈ {TEVd(n), SEVe(n), <TEVd(n)|SEVe(n)>, <SEVe(n)|TEVd(n)>}

then

PxtTRJw(n) = <PxtTRJw(n-1)|STEVw(0),STEVw(n)>

The preceding definition can be rewritten so as to evidence the space-time coordinate property of PxtTRJ and its value, as follows.

Definition 3b. PxtTRJ's, STEV's, and their properties

with

a) a) the conventions in Definitions 2, and 3a
b) xt = <<(x-sense, x-mag>,t> where x-sense := along-trajectory sense
c) PxtTRJ:Pxt:Vxt-v = <x.t> = <<x-sense-v, x-mag-v>, t>:= the value of xt
d) xt-change = <x-change, t-change> = <<x-change-sense, x-change-mag>, t-change> := xt-change property in terms of the properties of: spatial extension change (x-change) and temporal duration change (t-change), where x-change is a sense-magnitude pair
e) STEV:Pxt-change:Vxt-change-v = (dx,dt) = ((dx-sense, dx-mag),dt) = ((x-change-sense-v, x-change-mag-v),dt) := the change in the value of property xt at a certain STEV
f) PxtTRJw(n).Pxt.V(x(n),t(n)) := PxtTRJ "w" composed of n events and having the space time coordinate property "xt" whose value is the pair (x(n),t(n)) with x, t relative to the beginning of the trajectory at (x(0),t(0))

and x is the (arc-length) space coordinate along the trajectory

g) STEVw(n).Pxt-change.V(dx(n),dt(n)) := the n-th space time change event STEVw(n) along PxtTRJw(n), having property xt-change, with value (dx(n),dt(n)), where dx(n) or dt(n) may be nil

then

$$PxtTRJw(n).Pxt.V(x(n),t(n))$$
$$=$$
$$<PxtTRJw(n-1).Pxt.V(x(n-1),t(n-1))|STEVw(0).$$
$$Pxt-change.V((dx(0),dt(0)),$$
$$STEVw(n).Pxt-change.V((dx(n),dt(n))$$

The space time values associated with a PxtTRJ can be defined in terms of the space time change values associated with the constituent STEV's of the PxtTRJ as follows.

Definition 3c. xt values as a sequence of value changes
with
• Definitions 2 in Table II, and 3b
then

$$PxtTRJw(n):Pxt:V<x(n),t(n)>$$
$$=$$
$$(<x(n-1),t(n-1)>|<x(0),t(0)>, (dx(n),dt(n)))$$
$$:=$$
value $<x(n),t(n)>$ as a sequence of values $(dx(k),dt(k))$, for k $\in[0,n]$

For a correct interpretation of the above definition it is of essential importance to realize that the space coordinate property x corresponds to an arc-length along a trajectory—not to a *milage*. Also the x-change = <x-change-sense, x-change-mag> property is an arc-length change.

The milage property "xt-mi" of a PxtTRJ can be defined as follows.

Definition 3d. PxtTRJ's, STEV's, and their milage related properties

with

a) xt-mi = <x-mi,t>
    := xt-mi property in terms of the properties of: milage (x-mi) and temporal duration t

b) xt-mi-change = <x-mi-change, t-change>
    := xt-mi-change property in terms of the properties of: milage change (x-mi-change) and temporal duration change (t-change), where x-mi-change = x-change-mag

c) PxtTRJw(n).Pxt-mi.V(x-mi-v(n),t(n)) := PxtTRJ "w" composed of n events and having the space time coordinate property "xt-mi" whose value is the pair (x-mi-v(n),t(n)) with x-mi-v, t relative to the beginning of the trajectory at (x-mi-v(0),t(0)) and x-mi-v is the value of the space coordinate along the trajectory as a milage

d) STEVw(n).Pxt-mi-change.V(dx-mi,dt(n)) := the n-th space time change event STEVw(n) along PxtTRJw(n), having property xt-mi-change, with value (dx-mi(n),dt(n)), where dx-mi(n) or dt(n) may be nil. The value dx-mi can also be written x-mi-change-v.

then

$$\text{PxtTRJw(n).Pxt-mi.V(x-mi(n),t(n))}$$
$$=$$
$$\text{<PxtTRJw(n-1).Pxt-mi.V(x-mi(n-1),t(n-1))|STEVw(0).Pxt-}$$
$$\text{mi-change.V((dx-mi(0),dt(0)),}$$
$$\text{STEVw(n).Pxt-mi-change.V((dx-mi(n),dt(n))>}$$

Obviously, since x-mi has no property corresponding to sense of motion along the trajectory PxtTRJw, the milage value x-mi-v(n) can only increase relative to x-mi-v(n-1)—in contrast with the arc-length magnitude value x-mag-v(n), which can either increase or decrease relative to x-mag-v(n-1) depending on the sense value of x-v(n-1) and the sense value of dx-mi(n)

4. The Velocity property

In Mathematical Physics "instantaneous" velocity values are traditionally modeled as ratios of space coordinate change divided by time change, in the limit of infinitesimal time change. In order to formulate a corresponding Discrete Event Physics definition various conditions and conventions must be introduced:

- since, at least at the most general level of definitions, Discrete Event Physics does not assume mathematical values for properties, it is necessary to represent velocity values (and space-time-change-values) as pairs consisting of spatial change values and temporal change values
- for the same reason (and also because only models of actually existing entities are allowed), the idea of infinitesimal is not useful, but STEV's offer an alternative basis for a velocity definition, since they consist of an interaction (PA) followed by its immediate successor (with no intervening PA)

The preceding considerations are incorporated in the definitions in Table III, where use is also made of the Discrete Event Physics convention that a set specification by means of an event type with no accompanying symbol means all events of the stated type: {STEV}= (STEVx,∀x).

Table III. Basic definitions of instantaneous velocity (i-vel)

Expression	Meaning
with	
{PxtTRJ:STEV.Pi-vel.Viv}:= the set containing all PxtTRJ's with i-vel value "iv" (the definition set for that value)	
{PxtTRJ:STEV.Pi-vel.Viv}:inv := the invariant associated with the definition set for value "iv" of the i-vel property of a PxtTRJ	
then	
Definition 4a. General i-vel Definition	
{PxtTRJ:STEV.Pi-vel}:inv	the characteristic invariant of the set whose elements are PxtTRJ's having <u>some</u> i-vel value
{PxtTRJ:STEV. Pi-vel.V(iv) }:inv	the invariant associated with all PxtTRJ's having value iv, i.e. iv itself
Definition 4b. i-vel as a function	
F\i-vel(PxtTRJw:STEV)	a function associating an i-vel value with a STEV; F\i-vel(PxtTRJw:STEV) = {PxtTRJ:STEV. Pi-vel.V}:sub:inv := the invariant associated with a certain subset of the STEV's having the i-vel property and contained in some PxtTRJ

As per the following Definition 4c, instantaneous velocity values can be understood to constitute a sub-set of space-time-change events.

Definition 4c. Instantaneous velocity values as an invariant associated with a sub-set of all space-time-change events

with

PxtTRJw(n).STEVx(n).Pi-vel.Viv(n) := PxtTRJw(n) as a sequence of STEV's each of which has an i-vel value iv that is a certain function of space time change events

then

PxtTRJw(n):STEVw(n):Pi-vel:Viv(n)

=

{PxtTRJ.STEV.Pxt-change.V}:sub:inv

:=

i-vel value iv(n) is the invariant associated with a certain subset of all STEV's of all PxtTRJ's

As was done for space-time extension in Table II, the invariant (value iv) characterizing the subset of xt-change events having that velocity value may also be defined in terms of functions: this approach is presented in the following definition.

Definition 4d. i-vel values as function values

with

- xt-change values as defined in Definitions 3b
- F\i-vel:= a function mapping STEV's into velocity values "iv"
- F\xt-change := a function mapping STEV's into xt-change values (dx, dt)
- F\i-vel-from-xt-change := a function mapping space time change values (dx,dt) into velocity values "iv"

then

PxtTRJw(n):STEVw(n):Pi-vel:Viv

=

F\i-vel(PxtTRJw(n):STEVw(n))

=

F\i-vel-from-xt-change(F\xt-change(PxtTRJw(n):STEVw(n)))

Ideas from Mathematical Physics would suggest that function F\i-vel-from-xt-change is a Discrete Event Physics analogue of the derivative dx/dt.

According to the preceding definitions of a PxtTRJ, each of its constituent STEV's will have a certain instantaneous velocity value, which may vary from one STEV to the next. Two successive STEV's can thus combine to form an event having the property of velocity change (v-change) having values (dv,dt). Such events are called RTEV's, for roto-translations because they are the simplest case of such motions (revolution with translation as a limiting case). RTEV's can be used as primitive events in the definition of PxtTRJ's and their v-changes can be combined to define the velocity "coordinate" of the whole PxtTRJ, which relates the velocity at its end to that at its beginning. This is detailed in the following definition.

Definition 4e. a velocity value as a sequence of velocity change events

with

- RTEVx(n) = <STEVx(n),STEVx(n+1)> := a roto-translation event
- RTEVx.Pv-change.Vdv:= a velocity change (v-change) event, i.e. an RTEV with v-change value dv
- PxtTRJw(n)=<PxtTRJw(n-1)|RTEV(0),RTEV(n)> := a PxtTRJ as a sequence of RTEV's
- vel := the velocity property of a whole PxtTRJ
- v-sense=dx-sense := the sense value of the velocity at the end of PxtTRJw
- v-mag := the "magnitude" of the velocity value
- v= <v-sense, v-mag> := the velocity value of a whole PxtTRJ (see comment following definition)
- dv-sense := the sense of the velocity change associated with an RTEV
- dv-mag := the magnitude of the velocity change associated with an RTEV
- dv = <dv-sense,dv-mag> :=the value of function F\v-change, corresponding to property v-change of an RTEV

- $v(k) = PxtTRJw(k):Pvelocity:V$
- $dv(k) = RTEVx(k):PV\text{-}change:V$
- $dv(0)=v(0) :=$ a special velocity change corresponding to velocity 'initialization'

then

$$PxtTRJw(n):Pvel:Vv(n)=$$
$$<PxtTRJw(n\text{-}1):Pvel:Vv(n\text{-}1)|RTEVx(0):Pv\text{-}$$
$$change:Vdv(0),RTEVx(n):Pv\text{-}change:Vdv(n)>$$
$$=$$
$$v(n)=<v(n\text{-}1)|dv(0),dv(n)>$$
$$:=$$

v as an initial value dv(0) followed by a sequence of velocity changes

The velocity value of PxtTRJw(n) can thus be understood to be the sequence of velocity change values associated with its constituent RTEV's. As such it is the Discrete Event Physics analogue of the Mathematical Physics construct "dx/dt as a function of time (or more generally space-time.)" Of course, it is also—and more fundamentally—a function of primitive primary events (STEV's), and can also be understood as a function of the index of the successive STEV's along a trajectory. From this perspective velocity is an existent that contributes to explaining the meaning of space time in dynamical terms such as the motion induced extension of trajectories: a trajectory exists with a certain space time extension because its velocity value is consistent with such an extension, said consistency obviously being reciprocal.

5. The acceleration property

Instantaneous accelerations are understood to be a sub-set of the set of all velocity changes just as velocity changes were considered above to be a sub-set of all space time changes. Thus just like velocity change, instantaneous acceleration is considered to be a property of RTEV's, as per its general definitions given in the following table.

Table IV. Basic definitions of instantaneous acceleration

Expression	Meaning
with	
{PxtTRJ:RTEV.Pi-acc.Via}:= the set containing all PxtTRJ's with instantaneous acceleration (i-acc) value "ia" (the definition set for that value)	
{PxtTRJ:RTEV.Pi-acc.Via}:inv := the invariant associated with the definition set for value "ia" of the i-acc property of an PxtTRJ	
then	
Definition 5a. General i-acc Definition	
{PxtTRJ:RTEV. Pi-acc}:sub:inv	the characteristic invariant of the sub set of all RTEV's of all PxtTRJ's having <u>some</u> i-acc value
{PxtTRJ:RTEV.Pi-acc. Via}:sub:inv_ia	the invariant associated with all PxtTRJ's having value ia, i.e. ia itself
Definition 5b. i-acc as a function	
F\i-acc(PxtTRJw:RTEV)	a function associating an i-acc value with an RTEV,

In Discrete Event Physics instantaneous acceleration can be defined in terms of velocity change, much in the same way as instantaneous velocity was defined above in terms of space time change.

Definition 5c. Instantaneous acceleration (i-acc) values as a sub-set of all velocity change events

with

PxtTRJw(n).RTEVx(n).Pi-acc.Via(n) := PxtTRJw(n) as a sequence of RTEV's each of which has an i-acc value ia

then

PxtTRJw(n):RTEVw(n):Pi-acc:Via(n)
=
{PxtTRJ:RTEV:Pv-change:V}:sub_v:inv
:=
i-acc value ia(n) as the invariant associated with a certain subset of all RTEV's of all PxtTRJ's

As was done for instantaneous velocity values in Definition 4d, the invariant (value ia) characterizing the subset of xt-change events having acceleration value ia may also be defined in terms of a function: this approach is presented in the following definition.

Definition 5d. i-acc values as function values

with
- v-change values as defined in Definition 4e
- F\i-acc:=a function mapping RTEV's into instantaneous acceleration values "ia"
- F\v-change:=a function mapping RTEV's into v-change values (dv, dt)
- F\i-acc-from-v-change := a function mapping velocity change values (dv,dt) into acceleration values "ia"

then

PxtTRJw(n):RTEVw(n):Pi-acc:Via
=
F\i-acc(PxtTRJw(n):RTEVw(n))
=
F\i-acc-from-v-change(F\v-change(PxtTRJw(n):RTEVw(n)))

6. Operational definition of space time related properties

The preceding definitions of space time related properties showed how such concepts can be formalized in Discrete Event Physics, but added little insight into their physical nature. This deficiency is corrected, at least in part, in the following by adding operational criteria (relative to measurement strategies) to the previous definitions. The resulting definitions are suggested as prototypes, presumed to evidence <u>necessary</u> operations, which are schematized in the following Figure 1, where a process PCASb is having property values measured by another process PCASa in the latter's rest frame (PCASa is along the time axis t).

Figure 1. Measurement of space time properties such as inter-event intervals and velocities. A, B, C are vectors to points representing interactions (I/O relations); a, b, c are vectors whose elements are lines representing O-I relations; vector PCASb represents a space time process (STEV sequence) having <u>constant acceleration sense</u>; it contains STEVb(j) whose associated instantaneous velocity is being measured by the (<u>acceleration-free</u>) space time process represented by vector PNoAa; the sequence of STEVc(i) form another PNoA whereby the measurement is effectuated.

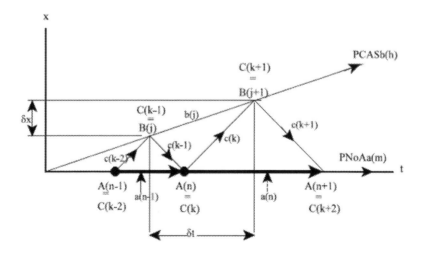

In order to formalize the measurement process in Figure 1, it is necessary to precisely define new event types: the type PNoA (no-acceleration) space time process and the STEVS space time event useful for taking into account different kinds of space time symmetry considerations.

In contrast with space time events of type STEV, whose fundamental characteristic is their anti symmetry (and consequent orientation), the new space time event type STEVS has the characteristic of being symmetric, in the sense that it implies both an orientation and its inverse.

For convenience in the following presentation it is also useful to introduce yet another space time event type STEVU, which indicates a space time event (STEV or STEVS) of (temporarily) unspecified type. The STEVS and STEVU types are defined formally in the following Definition 6a along with other process types, including the PNoA type.

Definition 6a. Space time primitive events and special processes

STEVab = <PAa,PAb>   := an ordered PA pair corresponding to an anti symmetric relation between PAa and PAb; PAb follows PAa unless a=b

STEVSab= (PAa,PAb)   := an unordered pair corresponding to a symmetric relation between PAa and PAb; unless a=b, PAb follows PAa and this implies the physical existence of the O-I relation in which PAa follows PAa, i.e. (PAa,PAb) = <PAa,PAb> and <PAb,PAa>

STEVUab= <PAa,PAb) or (PAa,PAb> = STEVab or STEVSab
:= a relation between PAa and PAb with unspecified symmetry properties;

PXT := a general space time process—it places no special constraints on the process

PCAS(h) = PXT(h).Pacc.Va := a PXT with acceleration value = "a" $\forall j \leq h$

PNoA(h) = PCAS(h).Pacc.Vnil := a PXT with acceleration value nil $\forall j \leq h$

PNoA(n) = <PNoA(n-1)|STEVU(0),STEVU(n)> where STEVU(0) = nil

For what follows it is very significant that the recursive definition of a PNoA specifies a nil event as the first event of a PNoA, in contrast with Definition 3a of a PxtTRJ, where no such condition is placed on the first event—this implies that

a PNoA can consist of the single event STEVU(1), since "nil" for STEVU(0) means "no event"

With respect to the relationship between PxtTRJ's and PNoA's there are two alternatives:

a) a PNoA is a special case of a PxtTRJ (symbol PXT in Definition 6a is identical with PxtTRJ) corresponding to a PxtTRJ with a nil zero-th event. This approach can present problems regarding the initialization of such processes—in previous discussions (e.g. Definition 3c) the zero-th event of a PxtTRJ was where initializations took place (of space and time coordinates, etc), but this function could be shifted to STEV(1)

b) PNoA's and PxtTRJ's are different in various ways, not just with regard to the nature of their zero-th event as being nil or not—so much different as to suggest that they correspond to special cases of a more general event type PXT. In this approach, it would seem advisable to place the explicit condition on PxtTRJ's that their zero-th events are never nil.

6.1 Space time invariant intervals

In special relativity space time interval changes (ds) are represented mathematically as

$$ds^2 = (c \times dt)^2 - dr^2$$

where c is the velocity of light and $dr^2 = dx^2 + dy^2 + dz^2$ and dt^2 are respectively the squares of the spatial distance and the time difference between points in space-time in any reference frame.

In the case that the interval corresponds to the trajectory of a system moving with velocity v, so that dr=vdt, then

$$ds^2 = (c^2 - v^2)dt^2$$

which reduces to

$$ds^2 = (c \times dt)^2$$

if v=0, i.e. in the rest frame of the system. Thus ds is the time interval in the system's rest frame (multiplied by c), and it has that value in all inertial reference frames.

In the following space time interval change will be referred to as property "sti-change", having value "sti-change-v" . The space time interval resulting from the concatenation of space time interval changes along a trajectory will be referred to as property "sti", having value "sti-v".

The measurement procedure modeled in Figure 1 is useful for defining different properties; it is formally described in the following definition with a specific application to the measurement of space time interval change values.

Definition 6b. Space time interval

with

a) Figure 1 for process and index correspondences and

a.1 sti := space time interval

a.2 sti-v := the value of sti

a.3 sti-change := the change in sti

a.4 sti-change-v := the value of sti-change

sti-change-v
=
<G\sti-change-sense(m,n,PNoAa, h,j,PCASb), F\
sti-change-mag(m,n,PNoA, h,j,PCASb)>
:=
the value of sti-change as a sense-magnitude pair of
functions
<G\sti-change-sense, F\sti-change-mag>

b) F\sti-change(m,n,PNoAa, h,j,PCASb) := the value of a function F\sti-change equaling the value of property sti-change of event PCASb(h):STEVb(j) as measured by (RTEVa(n-1) of) process PNoAa(n)

c) PCASb(j).Pacc.V(a-sense-v(j),a-mag-v(j)) = (line "b" in Figure 1)
and a-mag-v := acceleration magnitude value,
and acceleration sense value a-sense-v(j) is constant $\forall$ j≤n

d) PNoAa := a no-acceleration (inertial) space time "measurer-process" = line "a" (the time (t) axis) in Figure 1, where

d.1 PNoAa(m)=<PxtNoAa(m-1)|nil,RTEVa(m)>

d.2 RTEVa(n)=<STEVUa(n),STEVUa(n+1)>

d.3 STEVa(n)=<PA_A(n),PA_A(n+1)>

d.4   RTEVa(n-1):Pvel:Vv = RTEVa(n-1):STEVUa(n):Pi-vel:Viv := the velocity at the end of an RTEV (or, more generally, of a PART of a STEVU sequence, or even more generally of a PxtTRJ or a PNoA), has the same meaning as the instantaneous velocity of its final (end) STEVU.

e)   PCASc(K)=<PCASc(K-1)|nil,STEVc(K)> := a constant acceleration sense (see Definition 6a) space time process (the sequence of c(k)'s in Figure 1, for k≤K) connecting PNoAa to PCASb at the events (PA's) shown in the figure, that is:

e.1   PA_C(k)=PA_A(n), PA_C(k+1)=PA_B(j+1), etc as in the figure

e.2   STEVc(k-1)=<PCASb(h):STEVb(j):PA_B(j)), PxtNoA(m):STEVa(n):PA_A(n)>

e.3   STEVc(k)=<PA_C(k),PA_C(k+1)>
      =<PNoAa(m):STEVa(n):PA_A(n)),PCASb(h):STEVb(j):PA_B(j+1)>

e.4   STEVc(k+1)=<PA_C(k+1),PA_C(k+2)> =<PCASb(m):STEVb(j+1):PA_B(j+1),PNoAa(n):STEVa(n):PA_A(n+1))>

e.5   xt-change := xt-change as in Definition 3b where it = <x-change, t-change> and x-change is a change in arc-length

e.6   dx(k-1)=PCASc(K):RTEVc(k-1):STEVUc(k-1):Px-change:V
      dx(k)=PCASc(K):RTEVc(k-1):STEVUc(k):Px-change:V
      dt(k-1)=PCASc(K):RTEVc(k-1):STEVUc(k-1):Pt-change:V
      dt(k)=PCASc(K):RTEVc(k-1):STEVUc(k):Pt-change:V

f) <<dx(k-1),dx(k)>,(<dt(k-1),dt(k)>> :=
the ordered pair consisting of the arc-length of
RTEVc(k-1) followed by its duration, both in the
rest frame of PNoAa—
nb <dx(k-1),dx(k)> := dx(k-1) followed by dx(k)
<dt(k-1),dt(k)> := dt(k-1) followed by dt(k)

g) F\sti-change-from-xt-change(RTEV:Pxt-
change:V(dx,dt)) := a function that maps the
xt-change value of an RTEV into a sti-change-mag
value

then

PCASb(h):STEVb(j):Psti-change:Vsti-change-v

=

F\sti-change(m,n, PNoAa, h,j, PCASb)

=

F\sti-change-from-xt-change(<<dx(k-1),dx(k)>,(<dt(k-1),
dt(k)>>)

where

the correspondence between k, j and n is implicit in items
e.1 to e.4 and is explicit in Figure 1

It is important to recognize the importance of the nature of
xt-change as a (sense dependent) arc-length change as stated
in item e.5 of the above definition. Because of its dependence
on sense of motion along a trajectory, it can increase or
decrease the total arc-length x of the trajectory as the motion
proceeds from one STEV to the next along the trajectory.

As mentioned prior to the above definition, space
time intervals and interval changes have the important
characteristic of being invariant under change of a (non
accelerated) observer. This characteristic is formalized in the
following definition.

Definition 6c. Space time interval invariance
with
- Definitions 6a,b

then

$$\forall a,a', m'>n'$$
$$F\backslash sti\text{-}change(m',n', PNoAa', h,j, PCASb)$$
$$= F\backslash sti\text{-}change(m,n, PNoAa, h,j,PCASb)$$
$$:=$$

the value of the observed sti-change of a STEV is
independent of the observer

6.2 Ordered space time interval values

In (Delaney, 2004,2005 Tables 3.4 and 3.5) time (duration) values are defined as invariants associated with definition sets {PP.Pduration.V}, and value ordering is based on definition set ordering according to the criterion that {PP.Pp.Vv} is a successor of {PP.Pp.Vv'} if some elements of the former contain elements of the latter and no elements of the latter contain elements of the former.

When this condition is met the invariant (value) {PP. Pp.Vv}:inv is said to be related to {PP.Pp.Vv'}:inv by the "greater than" operator (.GT.). The sequential ordering of phenomena in time is then insured by defining the duration values of the sub-processes of a process as increasing monotonically with the process structure, i.e.

$$\forall t, PPx(t)\text{:}Pduration\text{:}Vx(t) \text{ .GT. } PPx(t-1)\text{:}Pduration\text{:}Vx(t-1) \quad (2)$$

The method of ordering definition sets and values is sufficiently general to be applicable to various kinds of existents and property values. Significantly, this general definition of value sequences is possible without reference to numbers.

Above, PxtTRJ's are understood to have the property of *space-time interval (sti),* also called *proper duration.* Assuming that the space-time interval of a PxtTRJ increases monotonically with the latter's structure leads to the following definition of space time interval ordering.

Definition 6d. Ordering of space-time intervals by the .GT. operator

with

$$PxtTRJw(n).Psti.Vsti-v(n)=$$
$$<PxtTRJw(n).Psti.Vsti-v(n-1)|STEVw(0).$$
$$Pxt-change.V(dx(0),dt(0)),$$
$$STEVw(n).Pxt-change.V(dx(n),dt(n))>$$

then

$$PxtTRJw(n):Psti:Vsti-v(n)$$
.is monotonic with.
$$PxtTRJw(n).Psti.Vsti-v(n)$$
so that
$$n > n'$$
$$\rightarrow$$
$$PxtTRJw(n):Psti:Vsti-v(n) .GT. PxtTRJw(n-1):Psti:Vsti-v(n')$$

6.3 Velocity

Here experimentation-based velocity definitions based on the measurement process in Figure 1 are proposed. The following Definition 7a extends Definition 6a by introducing new functions, F\iv-mag and F\iv-mag-from-xt-change, respectively mapping process pairs and xt-change values into instantaneous velocity values in analogy to how functions F\sti-change and F\sti-change-from-xt-change mapped such pairs and xt values into space time change magnitude values in the previous definition.

Definition 7a. instantaneous velocity magnitude value (experimental)

with

a)   Figure 1 and Definition 6a

b)   i-vel = ((i-vel-sense, i-vel-mag)) := the instantaneous velocity property as a sense-magnitude property pair

c)   iv = (iv-sense, iv-mag)) = ((i-vel-sense-v, i-vel-mag-v)) := the value of i-vel as a sense-magnitude pair

d)   iv = <G\iv(m,n,PNoA, h,j,PCASb), F\iv(m,n,PNoA, h,j,PCASb)>

:=

the value of i-vel as a sense-magnitude pair of functions <G\iv-sense, F\iv-mag>

e)   F\iv-mag(m,n,PNoA, h,j,PCASb) := the value of a function F\iv-mag equaling the iv-mag value of property i-vel-mag of event PCASb(h):STEVb(j) as measured by (RTEVa(n-1) of) process PNoAa(n)

f)   F\iv-mag-from-xt-change(RTEV:Pxt-change:V(dx,dt)) := a function that maps the xt-change value of an RTEV into an i-vel-mag value

then

PCASb(h):STEVb(j):Pi-vel:Viv

=

F:\iv-mag(m,n, PNoAa, h,j, PCASb)

=

F\iv-mag-from-xt-change(<<dx(k-1),dx(k)>,(<dt(k-1), dt(k)>>)

where

the correspondence between k, j and n is implicit in items e.1 to e.4 and is explicit in Figure 1

As stated above after Definition 4e, velocity is the Discrete Event Physics analogue of the Mathematical Physics concept construct "dx/dt as a function of time". Instantaneous velocity is the analogue of dx/dt at a specific time, or more exactly at a specific STEV (for F:\iv-mag) or at a specific xt-change (for F\iv-mag-from-xt-change).

Symmetry of instantaneous velocity values

It is significant that the procedure for measuring the instantaneous velocity magnitude value shown in Figure 1 and formalized in Definition 7a, evidences asymmetries between its constituent processes:

a) the <u>observed</u> process PCASb, whose associated instantaneous velocity magnitude value (at STEVb(j)) is being measured, is required to have constant (perhaps no) associated acceleration value whereas the <u>observer</u> process PNoAa is required to have no acceleration

b) the <u>observed</u> process PCASb participates in two interactions during the procedure, whereas the <u>observer</u> process PNoAa participates in three interactions.

Such asymmetries can have significant implications regarding the nature of velocity and the processes involved in its measurement. Supposing that the constant acceleration requirement for process "b" is strengthened to a <u>no acceleration</u> requirement (PCASb becomes PNoAb) so that it can be an inertial frame for the measurement of the instantaneous velocity of process PNoAa, then the inverse of an instantaneous velocity can be defined as follows.

Definition 7b. i-vel-mag value inverse

with

a) the contents of Definition 7a
b) inverse(<PNoAa,PCASb>) = <PCASb,PNoAa>
c) inverse(iv)=(inverse(iv-sense), inverse(iv-mag)))
d) PNoAb(h) = the special case of PCASb(h) where the constant acceleration sense of the latter is realized by its having no acceleration at all $\forall j < h$

then

inverse(F\iv-mag(m,n, PNoAa, h,j, PNoAb))

=

F\iv-mag(h,j,PNoAb, m,n,PNoAa)

Using Definitions 7a,b, in the special case where PCASb is actually PNoAb, the following hypothesis is meaningful.

Hypothesis 1: Velocity magnitude inversion symmetry

> iff   inertial process PNoAb contains an event (STEVU) that has an associated instantaneous velocitymagnitudevaluerelativetoaconstituent "measurer event" of inertial process PNoAa

> then   process PNoAa contains a STEVU that has the same associated instantaneous velocity magnitude value relative to a constituent "measurer event" of process PNoAb—i.e. the STEVU's in both processes are STEV's

Detailed formalization

with

a)   PNoAa(n) = <PNoAa(n-1)|nil,STEVUa(n)> := an inertial process containing 1 or more STEVUa's

b)   PNoAb(m).STEVUb(j).Pi-vel:=inertialprocess PNoAb containing STEVUb(j) with property i-vel

c)   PNoAa(m).RTEVa(n).STEVa(n) := inertial process PNoAa(m) containing RTEVa(n)=< STEVUa(n),STEVUa(n+1)> (see Figure 1), <u>which is a reference frame for the associated instantaneous velocity magnitude value of PNoAb(m):STEVb(j)</u>

d)   <PNoAa(m):RTEVa(n),PNoAb(h):STEVb(j)> :Pp:V := the value of property "p" of relation <PNoAa(m):RTEVa(n), PNoAb(h):STEVb(j)>, relating PNoAb(h):STEVb(j) to PNoAa(m):RTEVa(n) as the value's reference frame—see Definition 1b and its introductory text

then

$$\forall\ PNoAa(m):STEVa(n)$$

[$\exists$ <PNoAa(m):RTEVa(n), PNoAb(h):STEVb(j)> :
Pi-vel:Viv

$\rightarrow$

[$\exists$ <PNoAb(h):RTEVb(j), PNoAa(m):STEVa(n)> :
Pi-vel:Viv

$=$

]] <PNoAa(m):RTEVa(n), PNoAb(h):STEVb(j)> :
Pi-vel:Viv

$:=$

PNoAa and PNoAb consist of STEV's having the
same instantaneous velocity magnitude value
relative to each other (in each other's rest frame)

It is noteworthy how RTEVa(n) acts as the reference frame for the measurement of the instantaneous velocity of PCASb:STEVb(j)—see item c) in the above definition—implying that a STEV cannot act as a frame for such purposes.

Rejection of the hypothesis would imply that values of the property of instantaneous velocity magnitude are not symmetric in the hypothesized sense. Excluding considerations of measurement error, rejection of the hypothesis could be based on a single counter-example, which could be the result of an experimental observation or of a logical argument.

One such argument is suggested by special relativity according to which the instantaneous velocity of a massless system (e.g. a photon) cannot constitute a reference frame for the velocity magnitude of other systems. This idea could form a basis for rejecting the above Hypothesis 1, although it must be admitted that it is itself an hypothesis, not really the result of a logical argument (nor, obviously, of an observation).

Another argument for the rejection of Hypothesis 1 is based on the procedure for the measurement of instantaneous velocity illustrated in Figure 1 and formalized in Definition 7a.

Although the procedure requires only two interactions with the system whose instantaneous velocity is being observed, it requires three interactions with the observer system. Although the observed system could be a massless system (photon), it is less clear that it could be the observer system: what would the propagator be such that it is an emission in interaction PA_A(n-1) in Figure 1 and an absorption in interaction PA_A(n+1), and what would the nature of the interaction PA_A(n) be? Indeed it is not at all obvious that one photon can participate in three interactions.

Although the above arguments suggest rejection of Hypothesis 1 in the case in which one of the processes PNoAa or PNoAb corresponds to the motion of a massless system, they might actually not be relevant if certain conditions exist such as:

- there do not actually exist massless systems; they are an idealization
- massless systems do exist but they do not have the property of instantaneous velocity. Although they might exist in different places at different times they do not actually "move" between those places (they are never in between them, which might prove to be a valid criterion for defining the concept of velocity and even of motion itself.)

Realization of such conditions would avoid having to reject Hypothesis 1 and the inversion symmetry implied by the hypothesis could be understood as a defining characteristic of the velocity property.

Ordered instantaneous velocity values

An operational approach to the definition of *ordered* i-vel values is the following Definition 7c, which utilizes a particular feature of the trajectory PCASb—the possibility of velocity magnitude <u>change</u> along the trajectory while maintaining a fixed velocity sense so that velocity magnitude cannot <u>decrease</u>.

Definition 7c. Ordered instantaneous velocity magnitude values: .GE.

with

- the contents of Table I and Definition 7a
- iv-mag(j)=PCASb(h):STEVb(j):Pi-vel-mag:V
- iv-mag(i)=PCASb(h):STEVb(i):Pi-vel-mag:V
- j>i
- conventions whereby parentheses "[" and "]" respectively define beginning and end of statement sequences to be considered as logical units (with "]" at the <u>beginning</u> of the last statement in a sequence)
- .GE. := greater than or equal

then

$$PCASb(h):STEVb(j):Pi\text{-}vel\text{-}mag:Viv\text{-}mag(j)$$

$$.GE.$$

$$PCASb(h):STEVb(i):Pi\text{-}vel\text{-}mag:Viv\text{-}mag(i) \; \forall(i){<}j$$

$$\leftrightarrow$$

[PCASb(j) .contains. PCASb(i)

and ∄

   [PCASb(H).STEVb(J).Pi-vel-mag.Viv-mag(J)

and

   ]   PCASb(H).STEVb(I).Pi-vel-mag.Viv-mag(I), I<J

such that

  [iv-mag(J)=iv-mag(i), and iv-mag(I)=iv-mag(j)

and

    [∄ PCASb(H).STEVb(J).Pi-vel-mag.Viv-mag(J)

.contains

]   ]   ]   ∄ PCASb(H).STEVb(I).Pi-vel-mag.Viv-mag(I)

6.4 Velocity of propagation of interactions

The above Definitions 7a-c were concerned with the velocity values of the trajectory of an observed system PCASb (or PNoAb) relative to an observer trajectory PNoAa, with the observation mediated by a trajectory PCASc. The frame of reference of the observation is the rest frame of the observer, and therefor the trajectory of the latter has no spatial component—it is along the time axis. In Figure 1 it is obvious that not only PCASb has a velocity, but also PCASc has one—the *interaction propagation velocity* of the PNoAa-PCASb interaction.

The following Definitions 8a-c establish the condition that, in the context of such a mediated observation, the interaction propagation velocity i-vel associated with PCASc cannot be less than that of STEVb(j), essentially because <u>over the same time interval</u> associated with STEVb(j), PCASc follows a zig-zag spatial path between PNoAa and PCASb which is necessarily not shorter than the spatial extension of STEVb(j).

Definition 8a. Velocity of propagation of interaction from sub-events with velocity changes

with

Figure 1

RTEVc:Pipv := the interaction propagation velocity property of RTEVc

RTEVc:Pipv:Vipv-v : = the interaction propagation velocity value ipv-v of RTEVc

STEVc:Pipv-change := the interaction propagation velocity change property of RTEVc

STEVc:Pipv-change:Vipv-change-v : = the interaction propagation velocity change value ipv-change-v of STEVc

then

$$RTEVc(k-1).Pipv.Vipv(k-1)$$
$$=$$
$$<STEVc(k-1).Pipv\text{-}change.Vipv\text{-}change\text{-}v(k-1), STEVc(k).$$
$$Pipv\text{-}change.Vipv\text{-}change\text{-}v(k)>$$

Definition 8b. Velocity of propagation of interaction:
functional forms

with

Figure 1 especially for definition of index k

PCASc(K) := the trajectory used by PNoAa to observe
PCASb

PCASc(K):RTEVc(k-1) := <STEVc(k-1),STEVc(k)>

xt-change as per Definition 3

RTEVc(k-1):Px-change-mag:V
= <STEVc(k-1):Px-change-mag:V, STEVc(k):Px-
change-mag:V>
:= the spatial extension of RTEVc(k-1)

RTEVc(k-1):Pt-change:V <STEVc(k-1):Pt-
change:V,STEVc(k):Pt-change:V>
:= the duration of RTEVc(k-1)

RTEVc:Pipv : the interaction propagation velocity
property of RTEVc

RTEVc:Pipv:Vipv-v : the interaction propagation
velocity value ipv-v of RTEVc

F:\ipv := a function that maps an RTEV into an
interaction propagation velocity value

F:\ipv-from-xt-change := a function that maps an xt-
change-value into an interaction propagation velocity
value

RTEVc(k-1):Pxt-change:V = <STEVc(k-1):Pxt-change:V,
STEVc(k):Pxt-change:V> = F\xt-change(RTEVc(k-1)):=
the xt-change-value associated with RTEVc(k-1)

then

$$\text{RTEVc(k-1):Pipv:Vipv-v}$$
$$=$$
$$\text{F:\textbackslash ipv(RTEVc(k-1))}$$
$$=$$
$$\text{F:\textbackslash ipv-from-xt-change(F:\textbackslash xt-change(RTEVc(k-1)))}$$

The instantaneous velocities of PCASb(h):STEVb(j) and PCASc(k-1):RTEVc(k-1):STEVc(k) are compared in the following Definition 8c.

> Definition 8c. Constraint on instantaneous velocity vs interaction propagation velocity

with
- Figure 1, Definitions 6a, 8a
- ipv-change = <ipv-change-sense, ipv-change-mag> := interaction propagation velocity change as a sense-magnitude pair
- xt-change as per Definition 3b

then

$$PCASc(K):RTEVc(k-1):STEVc(k):Pipv-change-mag:V$$

$$.GE.$$

$$PCASb(h):STEVb(j):Pi-vel-mag:V$$

$$\leftarrow$$

[[PCASc(K):RTEVc(k-1):Px-change-mag:V

.GE.

]   PCASb(h):STEVb(j):Px-change-mag:V

and

[PCASc(K):RTEVc(k-1):Pt-change:V

.EQ.

]]   PCASb(h):STEVb(j):Pt-change:V

Recalling Hypothesis 1 and its accompanying commentary, the possibility that PCASc might not have the velocity property cannot be ignored. For example: what if PCASc is a model that is not isomorphic to a real existent (only its constituent STEV's are)? However even if this were true it is noteworthy that the comparison in Definition 8c is fundamentally one between xt coordinates, as all velocity comparisons are in an experimental context, and the comparison can be understood to refer to that level.

Definition 8c has significant implications in relation to Special Relativity. Indeed a basic tenet of that theory is that motions of bodies with a greater velocity than the "maximum velocity of propagation of interaction" is impossible (Landau and Lifshitz, The Classical Theory of Fields, 1951). Although a non infinite maximum (actual) velocity limit is to be expected, especially in Discrete Event Physics where infinities are excluded, the propagation of interaction velocity restriction is significantly more specific.

In contrast with its above characterization, the limiting velocity magnitude value is often also identified in Special Relativity with the velocity of propagation in vacuum of a "free wave" (i.e. one that propagates in an unrestricted space). Obviously this latter limiting velocity is quite different from the interaction propagation one—not only because they are logically different, but also because all interactions, even electromagnetic ones, seem to be necessarily understood to be confined to restricted spatial regions—unless, of course, they are non-local, in which case more criteria are needed to limit their confinement options.

The following table summarizes the various above cited considerations regarding the limitation of velocity magnitudes.

Table VI Limitation of velocity magnitude: issues and their possible resolutions

with	
	PM := a massive system := a trajectory (PxtTRJ) with the property of mass
	v := the instantaneous velocity magnitude value of a massive system
	c := the frame invariant (absolute) instantaneous velocity magnitude value of a free (spatially unbounded) massless system (e.g. photon) in vacuum
	c' := the frame invariant (absolute) space-time change value (dx,dt) of a free (spatially unbounded) massless system (e.g. photon) in vacuum
	w := the maximum velocity of propagation of interactions mediated by exchange of massless systems (propagators)
	w' := the maximum space time change value (dx,dt) of interactions mediated by exchange of massless systems (propagators)
	.LE. := less than or equal (not necessarily between numbers)
	.vs. := versus

then		
	Issue	Resolution
1	v .LE. c .vs. v .LE. w	v .LE. w ← (∄ free massless systems | v .LE .w .LE. c)

2	why a massless system cannot be a frame of reference (e.g. for the measurement of an instantaneous velocity magnitude property value) ?	a massless system cannot be a frame of reference <u>because</u> it cannot measure the instantaneous velocity magnitude property value of itself or another system by interacting with the latter system through the exchange of multiple propagators because it contains only two events (1 emission and 1 absorption)
3	v .LE. c .vs. v' .LE. c'	do comparison as v'.LE. c' because both massive and non massive systems have the xt-change property, whereas non massive systems do not have the property of instantaneous velocity
4	why is (dx, dt) a velocity for a massive system but not for a photon ?	it may be that (dx,dt) is local for a massive system and non local for a non massive system

7. Annotated summary

Sections 3, 4, and 5 respectively presented basic definitions of the properties of coordinates, velocity, and acceleration in the formalism of Discrete Event Physics. For each such property a hierarchy of sub-properties was defined, so as to evidence its various aspects: its temporal ones and its spatial (orientation and magnitude) ones. Each property had an associated property related to the change in its value, e.g. velocity-change for velocity. The property-change properties had the same hierarchical structure as the properties themselves. Each of the properties and their associated property-change properties were defined in terms of their underlying discrete events.

In Section 6 details were added to the property definitions in the preceding chapters on the basis of procedures proposed as being prototypical of their measurement. The idea of invariant space time interval was added and different connotations of the idea of velocity were considered, including that of "velocity of propagation of interactions". The treatment evidenced a symmetry in the velocity property that would seem essential to its definition, although an entity such as a (free) photon would not be expected to have it—suggesting that entities such as the latter may not have said property although they might have that of space time change. The definition of "velocity of propagation of interaction" evidenced why the value of said property is always greater than that of entities involved in the interaction—the basic tenet of Special Relativity.

An obvious limitation to the contents of Section 6 is the restriction to measurement in inertial frames, specifically by the no acceleration process PNoAa in Figure 1. This is of course the same limitation as that of Special Relativity, which is overcome in General Relativity. Actually, the results of Section 6 do not really seem to depend on the inertial frame restriction, i.e. generalization to non-inertial frames could be simply achieved by replacing symbol PNoAa by PCASa. This would be particularly desirable since it is not at all obvious that inertial frames (non accelerated motions) actually exist because of the pervasive presence of gravity.

The content of Section 6 also supports the explanations as to why physical space is three dimensional that were first published in (Delaney, 2004, 2005), where it was argued that, <u>in a dynamical theory of existence</u>, motions involving revolution or rotation are a necessary and sufficient condition for the existence of a physical three dimensional space. Indeed Figure 1 and Definition 6b and 7a suggest that the existence of revolution (as evident in the structure of trajectories PCASb and especially PCASc of Figure 1, and in the corresponding use of the RTEV structure in the cited definitions) is generally necessary for the measurement of space time coordinates and velocities, which, in the operational paradigm, implies its necessity for the existence of said properties.

4

Applicability of The Operational Paradigm

1. Introduction

The systematic organization of knowledge characteristic of scientific activities relies strongly on an appropriate selection and definition of basic concepts. An approach to definition of scientific concepts that is widely accepted, at least in principle, is Operationism, where concepts are defined in terms of operations, especially measurement operations.

One of the earliest and most influential exponents of this approach was P. Bridgman. In (Bridgman, 1927) he discusses the merits of operationism and points out certain difficulties associated with the approach. Among the many important ideas he presents, the following are especially relevant in the present context:

(2.1) the fundamental importance of (measurement) operations in defining physical quantities is elucidated in the statement *"the concept is synonymous with the corresponding set of operations"* (author's italics)

(2.2) difficulties with identifying such measurement sets, especially for the definition of fundamental quantities like distance and time at different (microscopic, intermediate, and astronomical) spatial and temporal scales

In (Delaney,1999) it was demonstrated that Operationism has limited applicability in Physics. The demonstration was based on a specific counterexample. Here the limitations of Operationism are extended to a considerable degree.

2. Measurement Procedures

In the following the above mentioned operation sets will be understood to be measurement procedures, i.e., operations to be performed in a specific order. Focus will be on the application of such procedures to the definition of properties and property values.

2.1 Measurement Procedure Requirements

(Delaney, 2005, Chapter 2, Table 2.1) lists the following characteristics of a measurement procedure, (MP) considered necessary for its being able to define a property.

Table I. Necessary characteristics of measurement procedures (MP's)

Characteristic	Meaning
procedural nature	a MP is a procedure and thus can have sub-procedures
uniqueness	a MP must execute a unique sequence of operations in any specific observational situation, so that it can define a unique concept
completeness	a MP must yield a result (property value) in all relevant observational situations
correctness	a MP must yield the correct result in all relevant observational situations
finiteness	a MP must yield a result in a finite time

Characteristic	Meaning
actual execution	a MP defines a property value inasmuch as it is actually executed
validity	a MP always yields a value for an entity having the property defined by the procedure and never yields a value for an entity which does not have that property

The properties of correctness, completeness and finiteness have a procedural nature; they seem to be self evident requirements. Actual execution implies understanding the meaning of a property/value by observing the execution of the measurement procedure which measures it, and not, for example, by studying a textual rendition of the procedure.

With reference to the "finiteness" property, "yielding a property value" is understood to include two distinct aspects: determining the value and "returning" the value, this latter term meaning the outputting of information identifying the value as the measurement result in an unambiguous and clearly recognizable form (in a finite time). If a procedure does not return a value under certain conditions, it will be said to "never return a value" (meaning never under those conditions).

The criteria presented in Table I are intended as being necessary conditions that a measurement procedure must respect. No claim is made that they exhaust all such conditions nor that in they constitute a sufficient condition, singly or together. Thus, at most, they can be considered to constitute an incomplete definition of "measurement procedure".

2.2 Measurement Procedure Structure

In (Delaney, 2005) a measurement procedure was understood to correspond to an activity (event type PA), corresponding to a notion of interaction general enough to include occurrence over an extended time interval. Here the just mentioned characterization of MP is generalized slightly: a MP is to be understood in the following as corresponding to a sequence of events (PA's), called a physical process (event type PP). A measurement procedure is a special case of a PP and the constituent PA's of a measurement procedure are measurement operations. The procedure is the same independent of what system executes it. Its inputs are those into its beginning event and its output is that out of its end event. The inputs correspond to influences coming from its environment, one such influence being the existent whose property value is to be measured. The output is the property value.

3. Measurement Procedure Formalization

The following is a definition of a measurement procedure, MP-MP-Q, for classifying existents as being procedures for the measurement of a property 'q' or not.

Definition 2.

I/O Specification for a measurement procedure, MP-MP-Q, for classifying a procedure PEy as being a procedure for the measurement of a property 'q' or not and, if PEy is such a procedure, for returning the value of q for a certain existent PEx

MP-MP-Q (PEy(PEx)) =

if

procedure PEy has property 'is or is not a measurement procedure for property q' with value 'is a measurement procedure for property q'

then if (PEx has property q)

then

return

1. PEy is a measurement procedure for property q
2. the value of q for process PEx

stop

else

return

PEx does not have property q

> stop

else

> return
>> PEy is not a measurement procedure for property 'is
>> or is not a measurement procedure for property q'
>
> stop

Being an I/O specification, Definition 2 says nothing about how MP-MP-Q works internally. As an aide to understanding, the following scenario is suggested:

> MP-MP-Q is performed by a measurement apparatus (which may include a person). The apparatus observes a process PPy and tries to ascertain if it is a measurement procedure (a MP-MP-Q) for a certain property 'q'. It would seem logical that, for MP-MP-Q to make such a judgement, PPy must be interacting with some physical existent (process or activity) PEx. The task of MP-MP-Q is to judge whether or not such an interaction corresponds to PPy measuring the value of the property 'q' of PEx.

4. Measurement Procedure Limitations

In (Delaney,1999) it was demonstrated that the hypothesis that all properties can be defined by means of measurement procedures can be rejected because it leads to logical contradictions. The demonstration explicitly evidenced such a contradiction for the property 'being a measurement procedure or not'.

Similarly, logical contradictions arise from hypothesizing that a measurement procedure, such as the MP-MP-Q procedure presented in Definition 2, can provide a means for defining the property of 'being a measurement process for property q'. Using the definitions

> F:DOMAIN := the set of all existents that can be inputs to function F
>
> F:RANGE := the set of all values that can be output values from function F.

F:Spec := an I/O specification of the function F.
F:value := the value of function F (the output it returns)

such a contradiction can be evidenced by using MP-MP-Q to define another procedure "PEz" having the following input-output function specification (definition).

Definition 3. I/O Specification for a procedure, PEz

PEz(PEy(PEx)) =
if (MP-MP-Q(PEy(PEx)) returns

1. PEy is a measurement procedure for property q
2. the value of q for process PEx

then 'never return a value'
else return the value of some property of some existent, such as PEy or PEx

With specific reference to Definition 3 in the special case that PEy=PEz, PEz returns a value if MP-MP-Q, as defined in Definition 2, says it should not and does not return a value if MP-MP-Q says it should. This is the same kind of result obtained in the above quoted proofs as to the non existence of a specific kind of measurement procedure. It implies that PEz does not exist which implies that MP-MP-Q does not exist.

For example, in the case in which q= the property of randomness, and PEx is a random process then two contradictions vitiating the required general validity of MP-MP-Q may arise:

• when and PEy=PEz is a measurement procedure for randomness PEy will not return the randomness value associated with PPx even though MP-MP-Q correctly says that it should do so

- conversely, if PEy=PEz is not a measurement procedure for randomness it will return a value as MP-MP-Q says it should not

5. Commentary

The preceding result as to the non-existence of a procedure capable of determining if an arbitrary process has a given property contradicts the basic tenet of Operationism, i.e. that any property of any physical entity can be defined by means of a measurement procedure. In other words, the result demonstrates that no property can be defined by means of a measurement procedure.

In such a context it is noteworthy that the contradiction arising from using the Operational approach to defining the property 'being a measurement procedure or not' as cited above at the beginning of Section 4, was shown, in (Delaney, 2005,Chapter 2), to be avoidable if the requirements on measurement procedures as stated in Table I are weakened so that such procedures are permitted to sometimes be inaccurate or to sometimes be non-committal (as to the nature of an input process). However interesting such observations might be on an abstract level it is difficult to imagine their practical significance—for this reason they are not repeated here.

Another way of avoiding the contradiction arising from using the Operational approach to defining the property 'being a measurement procedure or not' as that presented in Sections 3 and 4 above, would be to weaken the generality requirements on measurement procedures, assuming instead that

no measurement procedure can ever be executed more than once

so that, in particular, it can never refer to itself.

5

PROBABILITY

In the following a definition of probability relying on principles underlying Discrete Event Physics is presented. The relevant principles include objectivity, finiteness, and reference to only actually existing entities. Using the formalism of Discrete Event Physics:

- EV.Ppr := an event of type EV having property pr
- {EV.Ppr} the set of all events of type EV, anywhere in 'space-time', that have property pr
- prval(i) := the i-th value of property pr with $i \geq 0$
- {EV.Ppr.Vprval(i)} the set of all events of type EV that have property pr with value prval(i)
- m = the number of actually existing values of pr (anywhere in 'space-time')
- n(i) = the number of events having property prval(i)
- N = the sum of the n(i) for all prval(i) := the number of events having property pr

Using the above defined symbols

$$N = \sum_{i=1}^{m} n(i)$$

if $(prval(i) = n(i)/N)$ then $\sum_{i=1}^{m} (n(i)/N) = \sum_{i=1}^{m} (prval(i)) = 1$
and
if $n(i) > n(j)$ then the i-th prval occurs more frequently than the j-th one does

$$\text{implying that}$$
$$\text{prval}(i) > \text{prval}(j)$$

The above definition can be compared to the classic definition of frequency, where the $n(i)$ is limited to events in the past or present—here instead they can also include *future* events.

6

WAVES

1. Introduction

In the following, the most elementary wave constituents are referred to as being of type 'wave event' (WEV). Such events can be combined into sets and sequences. Particularly important ones are

a) sets of WEV's called wave *fronts* (events of type WFR)
b) sequences of WFR's called wave *trains* (type WTR).
c) sequences of WEV's called *rays* (type RAY)

One reason for the importance of waves in Physics is Huygens Principle, according to which each event in a wave front gives rise to *wavelets* that expand radially outward from it. This principle is a basis for models useful for explaining obscure phenomena :

1. how gratings can produce light bands on a screen due to interference of light traveling to the screen along non parallel directions from different slits on a grating
2. how a light source is able to illuminate points to which no straight line connection exists for a light particle (photon) to follow

Huygens Principle can also be formulated in terms of rays:

a) in one dimension a pair of RAY's with opposed spatial senses, called a *raylet*, emanates from each event along a RAY
b) in two dimensions a circularly symmetric set of outgoing RAY's emanates from each event along each RAY and is a special case of a RAY tree
c) in three dimensions a spherically symmetric set of outgoing RAY's emanates from each event along each RAY and is a special case of a spherically symmetric set of RAY trees

Correspondences exist between the above ideas and previously introduced representations of trajectories:

A) a sequence of RAY's having the properties of duration and spatial extension is an existent of event type PxtTRJ (space-time trajectory—(see Delaney 2005))
B) each RAY in such a sequence is a PART of the PxtTRJ
C) the elementary events (WEV's) in a RAY correspond to the elementary PxtTRJ events of type STEV (space-time event)
D) just as a PxtTRJ is a sequence of STEV's, a RAY sequence as a whole is itself a RAY corresponding to a sequence of WEV's

In consideration of the above items A-D, the following event type definition is introduced

Definition 1. Event type WxtTRJ
WxtTRJ := PxtTRJ corresponding to a WEV sequence
so that, recursively,
$$WxtTRJ(n) = <WxtTRJ(n-1),WEV(n)>$$

Detailed discussion of WxtTRJ's is organized as follows:

- Section 2 describes specific properties of such WxtTRJ's and their interactions

- Section 3 comments on difficulties inherent in the wave concept

2. WxtTRJ properties and interactions

Properties of WxtTRJ's are discussed in sub-section 2.1 and interactions between them is discussed in sub-section 2.2.

2.1 WxtTRJ properties

Space time position (xt) and *amplitude* (amp) are particularly important properties of WxtTRJ's.

A space time position value is the pair (x,t) where x is spatial position along a trajectory, relative to its beginning and t is duration relative to the trajectory's temporal origin.

An amplitude value (amp-v) is a pair

$$amp\text{-}v = (amp\text{-}s, amp\text{-}m)$$
$$where$$
$$amp\text{-}s := amplitude\ value\ sign$$
$$amp\text{-}m = amplitude\ value\ magnitude$$

It is convenient to define functions determining amp-s and amp-m values at space time positions along a trajectory; in the context of a certain trajectory having the properties of space time position and amplitude:

sf(x,t) := a function that maps <u>space time position</u> x,t into the <u>amplitude value sign</u>, amp-s, at that position, such that
$$amp\text{-}s(x,t) = sf(x,t)$$
and, at the i-th event along a trajectory,
$$amp\text{-}s(x(i),t(i)) = sf(x(i),t(i))$$

mf(x,t) := a function that maps the <u>space time position</u> x,t into the <u>amplitude value magnitude</u>, amp-m, at that position, such that
$$amp\text{-}m(x,t) = mf(x,t)$$
and, at the i-th event along a trajectory,

$$amp\text{-}m(x(i),t(i)) = mf(x(i),t(i))$$

The most important characteristic of amplitude values is their periodicity, defined as follows.

DEFINITION 2. AMPLITUDE VALUE PERIODICITY

with

inverse function : = a function that maps a value into its inverse

$\lambda :=$ wave length

$\tau :=$ time period

$(WxtTRJa(i).Pxt.V(x(i),t(i)))\text{:=}$event $WxtTRJa(i+n)$ having property xt with value $(x(i), t(i))$

$(WxtTRJa(i).Pxt.V(x(i),t(i)))\text{:Pamp:Vamp-}mf(x(i),t(i))\text{:=}$value amp-$mf(x(i),t(i))$ of property amp of the event $WxtTRJa(i+n)$ having property xt with value $(x(i), t(i))$

then

DEFINITION 2A. AMPLITUDE <u>MAGNITUDE</u> VALUE PERIODICITY

if($i+n \leq m$) then

$(WxtTRJa(i+n).Pxt.V(x(i)+n\lambda,t(i)+n\tau))\text{:Pamp:Vamp-}mf(x(i)+n\lambda,t(i)+n\tau)$

$=$

$(WxtTRJa(i).Pxt.V(x(i),t(i))\text{:Pamp:Vamp-}mf(x(i),t(i))$

and if($i-n \geq 0$) then

$(WxtTRJa(i+n).Pxt.V(x(i)-n\lambda,t(i)-n\tau))\text{:Pamp:Vamp-}mf(x(i)-n\lambda,t(i)-n\tau)$

$=$

$(WxtTRJa(i).Pxt.V(x(i),t(i))\text{:Pamp:Vamp-}mf(x(i),t(i))$

DEFINITION 2B. AMPLITUDE <u>SIGN</u> VALUE PERIODICITY

if($i+n \leq m$) then

$(WxtTRJa(i+n).Pxt.V(x(i)+n\lambda/2,t(i)+n\tau/2))\text{:Pamp:Vamp-}sf(x(i)+n\lambda/2,t(i)+n\tau/2)$

$=$

$inverse(WxtTRJa(i).Pxt.V(x(i),t(i))\text{:Pamp:Vamp-}sf(x(i),t(i))$)

and if(i-n ≥ 0) then

$$(\text{WxtTRJa}(i+n).\text{Pxt}.V(x(i)-n\lambda/2,t(i)-n\tau/2)):$$
$$\text{Pamp:Vamp-sf}(x(i)-n\lambda/2,t(i)-n\tau/2)$$
$$=$$
$$\text{inverse}(\text{WxtTRJa}(i).\text{Pxt}.V(x(i),t(i)):\text{Pamp:Vamp-}$$
$$\text{sf}(x(i),t(i)))$$

It is instructive to compare amplitude values to duration values as defined in Delaney, 2005, where duration was characterized as being a property of processes (event sequences), such that

- each sub process of a process has a duration value
- the duration values of the consecutive sub processes of a process are isomorphic with the sub process sequence
- the duration values are *unstructured*

Amplitude value sequences can also be characterized as being isomorphic with (abbreviation .iso w.) sub processes, e.g.

$$(\text{WxtTRJa}(i).\text{Pxt}.V(x(i),t(i)):\text{Pamp:V}) \text{ .iso w. } (\text{WxtTRJa}(i).$$
$$\text{Pxt}.V(x(i),t(i)).\text{Pamp}.V)$$
$$(\text{WxtTRJ}(i).\text{Pamp}.V(i) \text{ .iso w. inverse}(\text{WxtTRJ}(i):\text{Pamp}:V(i))$$

The major difference between duration values and amplitude values is how their structure affects how they vary with the sequence of sub processes with which they are isomorphic

- a <u>duration</u> value is simply a magnitude value and the duration value increase is <u>monotonic</u> with the increasing size of the consecutive sub processes of a process
- an <u>amplitude</u> value is *structured,* each amplitude value being a pair of values: a sign value and a magnitude value. The periodic variation of the amplitude <u>sign</u> value (as per Definition 2b above) induces variations in the amplitude <u>magnitude</u> value from increasing to decreasing values and vice versa.

2.2 Interactions

Two events, WEVb ∈ WxtTRJb and WEVc ∈ WxtTRJc, can combine to form a SUPER WEV, WEVa ∈ WxtTRJa, so that WEVb and WEVc are SUB events of WEVa. WxtTRJb and WxtTRJc are said to overlap or to interfere or to interact as WEVb and WEVc combine to form WEVa (or "coexist at" the space time coordinates of WEVa).

Amplitude values at SUB events (WEVb and WEVc) will combine to form an amplitude value at their SUPER event WEVa. Depending on the magnitude of the amplitude value at WEVa, compared to those at WEVb and WEVc, the interference of WxtTRJb and WxtTRJc is called either constructive or destructive, as explained in the following table.

Table IV Interactions

	expression	as existent: explanation	as mathematical model of existent: explanation
1	interference of WxtTRJb and WxtTRJc at WEVa	$\exists$ WEVa $\in$ WxtTRJa that CONTAINS both WEVb $\in$ WxtTRJb and WEVc $\in$ WxtTRJc	WEVa = WEVb + WEVc
2	constructive interference of WxtTRJb and WxtTRJc at event WEVa	interference of WxtTRJb and WxtTRJc at (WEVa $\in$ WxtTRJa) such that the amplitude magnitude value at event WEVa is greater than or equal to that at WEVb and that at WEVc	WEVa = WEVb+WEVc with sign(WEVa$\in$WxtTRJa) =sign(WEVb$\in$WxtTRJb) so that WEVa $\geq$ WEVb WEVa $\geq$ WEVc
3	destructive interference of WxtTRJb and WxtTRJc at event WEVa	interference of WxtTRJb and WxtTRJc at (WEVa $\in$ WxtTRJa) such that the amplitude magnitude value at event WEVa is less than both that at WEVb and that at WEVc	WEVa = WEVb+WEVc with sign(WEVa $\in$ WxtTRJa) = - sign(WEVb $\in$ WxtTRJb) so that WEVa < the greater of WEVb and WEVc

3. Commentary

The above characterization of Waves is similar to that in Mathematical Physics: its high level of abstraction permits various kinds of phenomena to be gathered together in one class—the class of periodic functions. Indeed the name 'periodic function' might be preferable to "wave". Although such recognition of commonalities is important in understanding the nature of existents, one must be careful to recognize the differences between the various kinds of waves, which can be considerable. A good example is the difference between waves whose very existence strongly depends on a medium through which they propagate, as compared to waves (like light waves and de Broglie waves) whose existence is totally independent of the presence of a medium.

An issue that was not considered at all above is the *locality* of the transmission of property values (e.g. energy values or probability values) from a source to a sink. Do the values propagate with the wave fronts or is the transmission non local, i.e. does it just disappear at its source and reappear, at a later time, at the sink? One thing that does not help at all in choosing between different hypotheses regarding wave (e.g. photon) properties and behaviors is that quantum mechanical lore prohibits observations that could provide crucial relevant information, for example observation of photon and de Broglie wave trajectories between their source and sink. Indeed it would seem preferable to define locality not in terms of space time coincidences, but in terms of ongoing existence within a super system that contains all events from a source to a sink.

7

PROPERTY MEANING, KNOWLEDGE AND UNDERSTANDING

1. Introduction

In the following, three concepts are defined: the *meaning* of properties, and the *knowledge* and *understanding* of such meaning. The definitions relate the concepts to objectively existing physical entities.

2. Existence of properties and property values

In Discrete Event Physics a property is defined as a set of property values, where a property value is defined as the set of all events having that value. Such a property value set structure is the "objective meaning" of the value, which is formalized in Discrete Event Physics as follows.

> Definition 1: meaning of 'objective property value'

with

> EV := the generic event type
> {EV.Pprop.VpropVal} := the set of all events of a certain type 'EV' and having property 'prop' with value 'propVal'

then

> the objective Meaning of value 'prop-val' := {EV.Pprop. Vprop-val}

which can also be expressed as
the objective Meaning of EV:Pprop:Vprop-val := {EV.
Pprop.Vprop-val}

In the following, the set of all property value sets, for all properties, is called the Universe. The subset of the Universe, all of whose constituent events have the properties "space-time" (xt) and energy-momentum (ep), is a good candidate for being called the Physical Universe.

2.1 Knowledge of property value meaning

As used in colloquial usage, the words 'knowledge' and 'knowing' have subjective aspects—they rely not just on the object about which something is known, but also on the subject who does the knowing, i.e. they regard a relation between a subject (the 'knower') and an object about which something is known.

Such knowledge can be therefore called 'subjective knowledge'. A different form of knowledge, regarding objectively existing entities, that does not rely on the knower is here called 'objective knowledge' and is defined, for property values, as follows.

Definition 2 : objective knowledge of property values
The Universe knows a property value := the Universe contains the events defining that property value.

Containment is not necessarily an immediate relation, e.g. it can contain a sequence of events the last of which is the event having the property value. Thus:
with
  a) <↓ a,b ↓> := a hierarchical relation from an event to a
     lower level event in an event hierarchy
  b) <↑ a,b ↑> := a hierarchical relation from an event to a
     higher level event in an event hierarchy
  c) HT := the event type of events corresponding to
     'hierarchical trajectories', i.e. sequences of events that

are hierarchically related in the sense that each contains its successor

d) $HT(n) = <\downarrow HT(n-1),\{EV(n).Pprop.VpropVal(n)\}\downarrow> := a$ sequence of hierarchically related events $EV(0).Pprop.VpropVal(0)$, . . . , $EV(k).Pprop.VpropVal(k)$, . . . , $EV(n).Pprop.VpropVal(n)$, where for all $k<n$, $EV(k-1).Pprop.VpropVal(k-1)$, is hierarchically superior to $EV(k).Pprop.VpropVal(k)$

e) $EV(0).Pprop.VpropVal(0)$ = the Universe

then

the objective knowledge of the meaning of value
$$EV(n):Pprop:Vprop\text{-}val(n)$$
$$:=$$
The Universe .CONTAINS. ($HT(n)$ tat .CONTAINS. $EV(n)$. Pprop.Vprop-val(n))

As defined in the above Definition 2, a hierarchical trajectory can have diverse interpretations as

a) a process of introspection whereby the Universe obtains knowledge or understanding of itself and its parts
b) a process whereby the Universe transports (gives) meaning to its parts

2.2 Understanding of property value meaning
In Discrete Event Physics a property value is the invariant associated with a set of discrete events; the actual existence of such an event set implies

a) the actual existence of the property value, which it defines and
b) the property value meaning, since the definition of a property value should imply its meaning

Defining a structured set as a set each of whose elements is related in a specific way (i.e. by a specific relation) to some other element (e.g. an element itself may be either a successor or a predecessor of any other one) then a set of events (or of event constituents) might form e.g.

- an unstructured event
- a relation between events
- a space-time sequence of events
- a hierarchical sequence of events, e.g. from the universe as a whole to an unstructured event—see the hierarchical sequence HT(n) defined above in Definition 2, item d.

Thus, with relation RELb(k) = <EVa(k),EVa(k+1)>, hierarchical event trajectory examples are:

1. HTa(n) = { EVa(0) , . . . ,EVa(n) }.Pprop1.Vpropval1, where EVa(0) = the Universe and propval1 corresponds to 'meaning of the trajectory, and in particular, of the relationship between the events EV(0) and EV(n)
2. HTb(n) = { <RELb(0)> , . . . <RELb(n)>}.Pprop2. Vpropval2 where propval2 might correspond, e.g., to either:

   a) *why* EVa(n) has propVal2, an obvious reason being because EV1(n) at the end of hierarchical trajectory HTa actually exists in the set {EV.Pprop.Vpropval} of all events having property 'prop' with property value 'propval'
   b) or *how* <RELb(n)> has propVal2, an obvious reason being because RELb(n) at the end of hierarchical trajectory HTb actually exists in the set {EV.Pprop. Vpropval} of all events having property 'prop' with property value 'propval'

Understanding why and how events and their properties come to have certain meanings is obviously of great interest, but it is difficult to come by: even after having been exposed to important clues and detailed explanations, people are very poor at recognizing or knowing meaning even partially. Examples abound where understanding of relativistic and quantum mechanical considerations are essential. A good example is provided by the property 'velocity'. People tend to think that objects 'have' velocities in some absolute sense—for example, early in their career students are taught that the earth has a velocity along a trajectory around the sun while the idea that it is the sun that moves around the earth is rejected. But later on they are told that motions are relative : which object moves around which other one depends on a choice of a reference frame, so that motion (velocity) can be understood to be a property of the relationship between the objects under consideration, not 'of' one or the other of them—all of which is quite counterintuitive to people.

4. Commentary

In the preceding three concepts have been defined:

1. the *meaning* of 'objective property value' as the set of actually existing events having that value
2. *objective knowledge* of a such an event property value as its attainment from events along a hierarchical trajectory that effectively 'transfer' such knowledge from the trajectory's highest event (the Universe)
3. *objective understanding* of a 'property value meaning' as knowledge about its genesis or persistence, with specific reference as to why and how event properties obtain their meanings from 'transferal' along hierarchical event trajectories that contain them

An advantage of using hierarchical event trajectories in concept definitions is avoidance of the difficulty in defining properties and their values in terms of experimental procedures, where output from an experiment is central; in the hierarchical approach output is replaced by *introspection*. In the above, definitions of meaning, knowledge, and understanding rely on reference to such trajectories, which can also be of great utility in the elimination of inconsistences arising in analyses of the Einstein, Rosen and Podolski experiment (Chapter 8).

8

ENERGY AND ENTROPY

1. Introduction

Thermodynamic energy and entropy are studied as properties of discrete event processes and process trees. Energy and entropy definitions are proposed and compared to each other and to discrete event definition of time duration. The following presentation is divided into 5 sections:

- Section 2: basic concepts and formalisms of Discrete Event Physics
- Section 3: energy is defined
- Section 4: entropy is defined
- Section 5: relations between duration, energy and entropy
- Section 6: summary

2. Concepts in Discrete Event Physics

The purpose of this section is to introduce basic concepts of Discrete Event Physics (DEP) in sufficient detail for the discussions of energy and entropy to be reasonably self contained. Definition of time duration is also summarized for following comparison to energy and entropy definitions.

2.1 Existent specification language

Existents are specified using linguistic forms similar to those of symbolic logic. The additional feature of *existent types* is also used. Table I displays a list of the basic forms and their definitions.

Table I. Basic notational conventions

Expression	Meaning
a:=b	assignment of the meaning of expression b to the meaning of expression a
a = b	a is the same as b
a|b	a or b
a∈b	a is an element of set b
∀	for all
∃	(there) exists
;	"such that" (in a set or structure definition, usually in conjunction with ∃ or ∀)
i∈[K,N]	"i" is a number greater than or equal to K and less than or equal to N
{def}:subsetX	subset "X" of the set defined by phrase "def";
{def}:subset	a subset of the set defined by phrase "def"
{def}:Subset	a proper subset of the set defined by phrase "def"
{def}:inv	the characteristic invariant associated with the elements of {def}
object	primary existent := material thing | event
TYPES	the set of object types
<a,b>	the ordered pair "a followed by b"
nil	"nothing", "empty", "zero", etc., as in "empty set", "empty sequence"
card(x)	the cardinality function that returns the number of constituent elements of x, where x is a set or a structure such as a system or complex event

Notational conventions for symbols representing primary and secondary existents at different levels of detail are shown in Table II.

Table II. Notational conventions for physical existents

Expression	Meaning
1. {T}	the set of all objects of type T ∈ TYPES
2. {T's}	the set of one or more objects of type T ∈ TYPES, i.e, {T}:Subset
3. Tx.Ba.Ez	the object "x" of type T∈TYPES, beginning at object "a", ending at "z"
4. {Tx}	the set consisting of the one object "x" of type T∈TYPES
5. Tx.Pp	the object "x" of type T∈TYPES having property p
6. {T.Pp}	the set of all objects of type T∈TYPES having property p
7. Tx.Pp.Vv	the object "x" of type T∈TYPES having property p with value v
8. T.Pp.Vv	an object of type T∈TYPES having property p with value v
9. Tx(i).Pp.Vv	the i-th object (of type T∈TYPES) in list "x" of such objects, having property p with value v
10. T(i).Pp.Vv	the i-th object of type T∈TYPES in some unspecified list of such objects having property p with value v
11. {T.Pp.Vv}	the set of all Tx.Pp.Vv, for all "x" := the definition set for value v
12. {T's.Pp.Vv}	{T.Pp.Vv}:Subset
13. Tx:Sy	the object "y" of type S∈TYPES as a <u>constituent</u> of object Tx (read Sy of Tx)

14. {Tx:S}	the set consisting of all Tx:Sy, for all y
15. {Tx:sub} = {Tx:subT}	{Tx:S} with S=T, S:B=Tx:B.
16. {Tx:S's}	{Tx:S}:Subset
17. Tx:Pp:V	a value of property p of object x of type T
18. Tx:Pp:Vz	the value z of property p of object x of type T
19. Tx:Sy:Pp:V	the value of property p of object y of type S in the context of object x of type T with S,T ∈TYPES
20. Tx:Sy:Pp:Vz	the value z of property p of object y of type S in the context of object x of type T with S,T∈TYPES
21. {Tx:Sy:Pp:V}	the set of all values Tx:Sy:Pp:Vz, for all z

In Table II the distinction between the symbols "."and ":", as in Tx.Pp.Vv and Tx:Pp:Vv, is of such importance in what follows as to merit additional attention. In general "."means "having"or "with", so Tx.Pp.Vv represents the type T object "x" having property "p"with value "v". On the other hand ":" means "of", or "in", so Tx:Pp:Vv means value "v"of property "p"of the type T object "x" and Tx:Sy, means Sx as some sort of *constituent* of Tx. What sort of constituent is intended depends on whether S=T or not. For example, supposing that S=T="set"then Tx:Sy would stand for the subset y of set x. On the other hand if T=set and S is not a set then Tx:Sy means that y is a type S element of set x. Note that expressions containing ":" are best read from right to left.

More complex expressions are also possible. For example, Tx:Sy:Pp:Vv means value "v"of property "p"of the type S existent "y" where y is a constituent of type T object "x". In contrast, (Tx.Pp.Vv):Sy signifies type S constituent "y"of the type T object "x where "x" has property "p" with value "v". Note the use of parentheses for ambiguity resolution.

The use of types without reference to specific objects has special significance. Thus {T} means all objects of type T and Tx:Pp:V signifies a value of property "p" without specifying which value.

Also T.Pp.Vv signifies an object having property p with value v, without specifying which object and {T.Pp.Vv} means all such objects. To refer to some objects of type T without specifying which ones, the form T's is used, as in {T's} which thus signifies a non-empty set of objects of type T.

The use of ":" with reference to sets is also very important, thus {T}:subset (more simply, {T}:sub) is a subset of {T}, {T.Pp}:subset is a subset of all objects with property "p" and {T.Pp.Vv}:inv is an invariant associated with all objects of type T having property p with value v, one such invariant being, of course, "having: type T, property p and value v".

2.2 Definition specification language

Special linguistic conventions for use in definitions are displayed in Tables III and IV respectively for definitions as meaning assignments and as conditions or hypotheses.

Table III. Basic formalism for definitions

Expression	Meaning
with	
A) prefix(1)\prefix(2)\ . . . \prefix(N)\concept B) prefix(1)\prefix(2)\ . . . \prefix(N)\concept:value C) PPx = <PPy|EVy,EVz>	
then	
concept	the idea being defined (in (A), typically a property or a relation)
concept:value	the value of a concept (property or a relation)

prefixes(optional)	are intended to represent the chain of key ideas leading to the definition in a very synthetic way; from the most general idea (prefix(1)) to the most specific one(prefix(N)).
Expression	Meaning
=	recursively equal (as in C), in that it applies also to PPy on its right
|	"or" (in C)
,	"followed by" (in C)
[a line of text zero or more lines of text] a line of text	multiple lines of text to be considered as a logical unit, typically as one of the two expressions related by a binary logical operator. The first line of the unit begins with "[" and the last line begins with "]". Such units may be imbedded within other ones; to enhance readability corresponding symbols "[" and "]" are vertically aligned.

Table IV. Formalism for definitions asserting conditions or hypotheses concerning meaning

with	
both a and b = statements that can: — contain conditional definitions — contain phrases connected by logical operators such as: and, or, | ("or"), not, ¬ (not)	
then	
a → b	a implies b, i.e., b is true if a is true. Can also mean that a is a generalization of b

a ← b	b implies a, i.e., a is true if b is true. Can also mean that b is a generalization of a
a↔b	a is true if and only if b is true; i.e., a and b imply each other
if a then b	same as a → b
y ← (∀x;b)	for y to be true it is sufficient that b is true for all x
y → (∀x;b)	for y to be true it is necessary that b is true for all x
y ↔ (∀x;b)	for y to be true it is necessary and sufficient that b is true for all x

To enhance understanding of later material, certain aspects of the tables and their contents deserve particular attention:

- the with-then clause at the table beginning; such clauses will be used extensively in tables and in definitions for the specification of information useful for understanding following content
- the recursive interpretation foreseen for the equality sign "=", which will be used extensively in definitions of primary existents
- the use of the square parentheses [,] to delimit multi-line expressions that must be understood to constitute conceptual units, e.g. one of the two terms being related by a binary operator, such as "and". This is similar to the use of special symbols in programming languages to delimit "blocks".

2.3. Events and processes
This sub section defines DEP concepts important for following definitions of energy and entropy:

- events and processes
- structures involving processes
- the process duration property

2.3.1 Events

Events (EV's) are defined as primary existents involving change, for example the emission of a particle by a radioactive source. Within that general conception, their definition is here refined as follows:

event := a (possibly structured) set of sub-events or
happenings

In this context, "happening" is to be considered as primitive. A set of one or more happenings constitutes a "simple" or "unstructured" event and a set of several events is referred to as a "complex" or structured event. This usage of the term "event" is commonplace in treatments of probability and statistics and differs from the simpler use of the same term in relativity theory, where it designates a point in space-time.

A *structured event* is a set of sub-events or happenings interrelated in a certain way. So interpreted, the idea of an event is more general than that of process (PP). That is, a PP is a special case of an event.

2.3.2 Processes as behaviors and activities

Using the ideas of *input* and *output* it is possible to classify processes into *activities* and *behaviors*. An activity (type PA event) is a (time) sequence of *I/O event's* called activity events (AEV's, which include the idea of interaction) that transform input's (causal factors) into outputs (effects). An O-I relation is an ordered pair $BEV(i)=<AEVx(i),AEVx(i+1)>$, which means it is an output of $AEVx(i)$ that is input to $AEVx(i+1)$. A behavior (event type PB) is a sequence of O-I relations. As examples: a differential equation like Newton's Law is a mathematical model of an activity (PA); the solution of such an equation as a function of time, for specific initial input values, models a behavior. PA's can also be characterized as ongoing activities (I/O relation's) consisting (hierarchically) of sub-PA's. Similarly, PB's are ongoing behaviors (O-I's = $<PAx(i),PAx(i+1)>$) that are hierarchies of sub-PB's.

The models in terms of I/O relations and O-I relations formalize the causality principle ; they can be unified by defining processes (PP's) as

$$PPx(t) = <PPx(t-1)|EVz(0),EVz(t)>$$
where
[PP=PA|PB
and
] EV:=a general simple event, which is an AEV for activities and a BEV for behaviors

The above recursive definition of a PP can be alternatively expressed as

$$\text{if } t>1 \text{ then } PPx(t) = <PPx(t-1),EVz(t)>$$
$$\text{if } t=1 \text{ then } PPx(1) = <EVz(0),EVz(1)>$$

The various ideas presented above with relative to PP's and their internal structure are synthesized in the following Table V which also introduces some new notational conventions.

Table V. PP's and their internal hierarchical-sequential structure

Expression	Meaning
1. PP	event type for processes
2. EV	PP primitive events
3.a) Recursive definition for PP's	PPx= <PPy|EVa,EVz>
3.b) PP recursive definition with integer valued indices; template	PPx(t) = <PPx(t-1)|EVz(0),EVz(t)>; PP(t) = <PP(t-1)|EV(0),EV(t)>
4. PPx(t).Bz(0)	PPx(t) beginning at EVz(0)
5. PPx(t).Bz(0) .is the immediate successor of. PPx(t-1).Bz(0) and Px(t-1).Bz(0) .is the immediate predecessor of. PPx(t).Bz(0)	PPx(t).Bz(0)=<PPx(t-1). Bz(0),EVz(t)>
6. PPx(t-1).Bz(0) .is a predecessor of. PPx(t).Bz(0)	PPx(t-1).Bz(0) .is the immediate predecessor of. PPx(t).Bz(0) .or. PPx(t-1).Bz(0) .is a predecessor of a predecessor of PPx(t).Bz(0)
7. PPx(t).Bz(0) .is a successor of. PPx(t-1).Bz(0)	PPx(t).Bz(0).is the immediate successor of. PPx(t-1).Bz(0) .or. PPx(t).Bz(0) .is a successor of. a successor of PPx(t-1).Bz(0)
8. PPx .contains. PPy and PPy .is contained in. PPx	PPx .is a successor of. PPy and PPy .is a predecessor of. PPx

9. PPy .is a sub-PP of. PPx and PPx .is a super-PP of. PPy	PPy .is a predecessor of. PPx
10. PPx .is a super-PP of. PPy	PPx .is a successor of. PPy
11. EVz(t) .is the immediate successor of. EVz(t-1) and EVz(t-1) .is the immediate predecessor of. EVz(t)	PPx(t).Bz(0)=<PPx(t-1). Bz(0),EVz(t)>
12. EVz(j) .is a predecessor of. EVz(k)	EVz(j) .is an immediate predecessor of. EVz(k) or EVz(j) .is a predecessor of. a predecessor. of EVz(k)
13. EVz(k) .is a successor of. EVz(j)	EVz(k) .is an immediate successor of. EVz(j) or EVz(k) .is a successor of. a successor of EVz(j)

2.4 Process duration

PP's often have the property of duration, in which case they are referred to below as duration processes. Duration values are defined (Delaney, 2004, Tables 3.4 and 3.5) as invariants associated with definition sets {PP.Pduration.V}, and value ordering is based on definition set ordering according to the criterion that {PP.Pp.Vv} is a successor of {PP.Pp.Vv'} if some elements of the former contain elements of the latter and no elements of the latter contain elements of the former. When this condition is met the invariant (value) {PP.Pp.Vv}:inv is said to be related to {PP.Pp.Vv'}:inv by the "greater than" operator (.GT.). The sequential ordering of phenomena in time is then insured by defining the duration values of the sub-processes of a process as increasing monotonically with the process structure, i.e.

$$\forall t, \; PPx(t):Pduration:Vx(t) \; .GT. \; PPx(t-1):Pduration:Vx(t-1) \quad (1)$$

The method of ordering definition sets and values is sufficiently general to be applicable to various kinds of existents and property values. Significantly, this general definition of value sequences is possible without reference to numbers.

2.5. Process hierarchies

A PP can also participate in a second kind of hierarchical relation, called a process hierarchy (PPH). A PPH is a set of process trees (PPT's). Each node in a tree corresponds to a PP that CONTAINS a set of SUB-PP's that are understood to correspond to concurrent processes (CONTAINS and SUB are capitalized to distinguish them from the above "contains" and "sub" relations.) The hierarchical relation IS CONTAINED IN exists between a PP and the SUPER-PP that CONTAINS it; the relations between the SUB-PP's of a specific PP are neither hierarchical nor sequential: they are causal (mediated by O-I relations). An example of such a process is a gas whose constituent molecules coexist and repeatedly participate in (energy exchanging) interactions with each other. The relation between a PP and its SUB-PP's can be recursively defined as: PPx={PP's}, meaning that PPx is a set of interacting PP's and each PP it CONTAINS also constitutes such a set of interacting PP's.

The preceding ideas relevant to process hierarchies are assembled for convenient reference in Table VI.

Table VI. Structural relations in PP hierarchies

Expression	Meaning
1. PPx={PP's}	a PP hierarchy, i.e., PP "x", as a set of SUB-PP's unified by interactions
2. PPx .CONTAINS. PPy and PPx .is a SUCCESSOR of. PPY	PPx={PPy,*}, i.e., PPx is a set of PP's that includes PP
3. PPy=PPx:SUBy =PPx:SUB-PPy and PPx=PPy:SUPERx=PPy:SUPER-PP	PPx = {PPy,*}, i.e., PPy is a SUB-PP of PPx and PPx is a SUPER-PP of PPy; where abbreviated forms (without "-PP") can be used when this does not cause ambiguity

2.5.1 Linguistic conventions

To minimize ambiguity, it is often important to distinguish special cases of sub, super, SUB and SUPER-PP's:

- the first super-PP of a PP is called the latter's "immediate" super-PP, or immediate successor
- the first sub-PP of a PP is called the latter's "immediate" sub-PP, or immediate predecessor
- the first SUPER-PP of a PP is called the latter's "immediate" SUPER-PP, or immediate SUCCESSOR
- any PP in the first set of SUB-PP's of a PP is called an "immediate" SUB-PP of that PP, or one of its immediate PREDECESSOR's

Such distinctions already appear in Table V above. They are collected together in the following tables, along with abbreviations for them, in order to clearly evidence their interrelationships, in particular which are "immediate versions" of others.

Table VII. Abbreviations for often used expressions

Expression	Abbreviation
immediate successor	i-successor
immediate predecessor	i-predecessor
immediate sub	i-sub
Expression	Abbreviation
immediate super	i-super
immediate SUCCESSOR	iSUCCESSOR
immediate SUPER	iSUPER
immediate PREDECESSOR	iPREDECESSOR
immediate SUB	iSUB

2.6 Process Trees

To be able to contextualize a PP, such as PPx as defined in Section 2.3.2, which may itself be a SUB-PP of another PP, it is useful to introduce the process tree (PPT) structure, i.e. a hierarchical total organizing of SUB-PP's. Letting $PPx(h)(t)$ be a process with sub-PP's $PPx(h)(k) = PPx(h)(t)$:sub's with $k<t$, then for fixed t and process $PPx(h)(t)$

$$PPTx(h)(t) = \{PPx(h)(t):SUB\}:HTO.$$

For each t value, $PPTx(h)$ is a hierarchy extending from a level corresponding to a PP having no SUPER-PP, down through a sequence of levels to a lowest level whose elements are all PP's having no SUB-PP's. The hierarchical index "h" in $PPx(h)(t)$ has the purpose of specifying the hierarchical collocation of PP's within $PPTx(h)(t)$: its value should be larger for $PPx(h)(t)$ than it is for any of the latter's SUB-PP's and is generally not simply related to the number of SUB-PP's of $PPx(h)(t)$.

The time index "t" has the same role as it had for PP's, except that here it (also) specifies the sequencing of whole trees $PPTx(h)$. That is, in $PPTx(h)(t)$ and $PPx(h)(t)$, the time index t does not vary with h; all the PP's in PPTx(h)(t) have the same number (t) of sub-PP's and of primitive events (EV's). This is a significant simplification of allowing t to be a general function of h; it can be rationalized by assuming the t value of the all constituent PP's of a PPT to be <u>inherited</u> from their iSUPER-PP. Said simplification is assumed throughout the following.

PPx(h)(t) is the <u>root node</u> of PPTx(h)(t) and the PPx(h) (t):SUB's that have no SUB-PP's are <u>leaf nodes</u> of PPTx(h) (t). Each PPx(j)(t)= PPx(h)(t):SUBx(j)(t) that is not a leaf node is the root node of another tree PPTx(j)(t), this latter being a sub-tree (SUB-PPT) of PPTx(h)(t) and an immediate sub-tree (iSUB-PPT) of PPTx(h)(t) if PPx(j)(t) is an iSUB-PP of PPx(h)(t). Here SUB-PPT (not sub-PPT) is used because not all SUB-PPT's of PPTx(h)(t) are hierarchically related <u>to each other</u> by the organizing of the latter (the tree organizing <u>of PP's</u> is not sequential).

The simplest example of a PPT corresponds to the situation in which two PP's, PPz(0) and PPzz(0) both have the same iSUPER-PP, PPzzz(1); PPzzz(1) is the root node that branches down into the two leaf nodes PPz(0) and PPzz(0). A general PPT definition is:

Definition 1. Process tree (PPT) as a hierarchy of sub-trees with
  a) PPx(h)(t) := the root node of tree PPTx(h)(t), formed by the interaction of its SUB-PP's
  b) PPx(k)(t) := a SUB-PP of PPx(h)(t) in the context of tree PPTx(h)(t), formed from interactions among the SUB-PP's in the tree PPT(k)(t) of which PPx(k)(t) is the root-PP
then
$$PPTx(h)(t)=<\{(PPTx(k)(t)|PPx(k)(t)),\forall x(k);PPx(k)(t)$$
$$=PPx(h)(t):iSUB\},PPx(h)(t)>$$
which evidences the 1-to-1 relationship between PPx(h)(t) and PPTx(h)(t), i.e.
PPx(h)(t) .is isomorphic with. PPTx(h)(t)

The recursion ends when the set corresponding to the first element in the ordered pair on its right hand side contains only leaf nodes; it assumes no convention for the choice of values of indices "t" and "h", except of course that they must be a unique pair for all PPx(h)(t) in a tree.

3. Energy

Here energy is studied as a property of primary existents, in close analogy with the above described treatment of duration as a property of duration processes and process trees. The most important characteristic of energy is its conservation. In the present paradigm, energy conservation means that if the energy value of a process PPx changes then that PP IS CONTAINED IN another PP, PPz=PPx:SUPER, whose energy value does not change. This implies that PPz CONTAINS PP's with both increasing and decreasing energy values. Thus energy value changes have a magnitude and a sense (increasing or decreasing). Describing PP's in terms of inputs and outputs connecting a sequence of PA's, PAx(i), for i∈[1,I], leads to the idea of energy flow: decreasing energy value in PAx(1) followed by increasing energy value in PAx(2), followed by decreasing energy value of PAx(2) and increasing energy value of PAx(3), etc., can be conceptualized as the flow of energy through space, from PAx(1) to PAx(2) to PAx(3), etc. The preceding description obviously does not imply that any of the flowing energy ever actually exists in the space between the various PAx(i)—that requires a very strong continuum hypothesis.

Saying that energy is conserved does not of course say what energy is. This problem will be affronted in the following by characterizing energy in terms of the primary existents of which it is a property. Specifically, energy will be considered as a property of processes, and not of systems (PS's) as static matter conglomerates. This is consistent with a totally dynamic perspective in which PS's are considered to be processes consisting of sets of SUB-processes whose interactions cause them to be perceived as maintaining a limited spatial extension while they endure over time (Delaney, 2004, Chapt. 4, Sect. 2.6).

A process having the energy property will be called an "energy process"; such processes are distinguished by the type symbol EP. For greater generality, they are not defined a-priori to be special cases of the above mentioned processes (PP) having the duration property.

3.1 Energy property definition

EP's are understood to be complex events whose constituents are sub-EP's and primitive events of type EEV. Each EP is formed by concatenating an EEV onto the end of its immediate sub-EP, which leads to the following recursive definition of such processes.

Definition 2. Energy Processes

$$EPx(t) = <EPx(t-1)|EEVq(0),EEVq(t)>$$

Basic energy definitions in terms of EP processes are presented in the following Table VIII using the formalism outlined in Table IV.

Table VIII. Basic definitions of energy in terms of energy processes

Expression	Meaning
with	
{EP.Penergy.Ve}:= the set containing all EP's with energy value "e" (the definition set for that value)	
{EP.Penergy.Ve}:inv := the invariant associated with the definition set for value "e" of the energy property of an EP	
then	

Definition 3. General Energy Definition	
energy	{EP.Penergy}:inv, i.e., the characteristic invariant of the set whose elements are EP's having some energy value
energy:value	{EP.Penergy.Venergy:value}:inv, i.e., an energy value "energy:value"
Definition 4. Energy as a function	
F\energy	a function associating a value "F\energy:value", with an EP, i.e., ∀EPx.Penergy, EPx:F\energy:value=F\energy(EPx)
F\energy:value	the characteristic invariant of the set of EP's with which F\energy associates the specific energy value "F\energy:value", {EPx; EPx:F\energy:value= F\energy(EPx)}:inv

The general definition defines energy in terms of the characteristic invariants of sets of EP's having the energy property. Considered as a function, energy maps an EP into a unique energy value.

3.1.1 Energy value ordering

More detailed energy definitions can be obtained by refining the idea of F\energy so as to consider specific kinds of such functions. An interesting example would be a function that would evidence the principle of energy conservation. The definition of such a function relies on being able to compare energy values. For such purposes the general method for defining and ordering values described above in Section 2.4 will be used. Thus if the some elements of the energy value definition set {EP.Penergy.Vv} contain elements of the definition set {EP.Penergy.Vv'} and no elements of the latter contain elements of the former then the energy values are related by the .GT. operator:

{EP.Penergy.Vv}:inv .GT.{EP.Penergy.Vv'}:inv

The treatment of energy will however require defining operators expressing ordering relations between property values other than just .GT. This is done in the following Table IX.

Table IX. Operators relating property values

Expression	Meaning
PPx:Pp:Vu.OP. PPy:Pp:Vv	value u of property p of PPx is related to the value v of p for PPy by the operator .OP.
Operators	
.EQ.	equal to
.NE.	not .EQ.
.GT.	greater than
.GE.	greater than or equal, i.e., .GT|EQ.
.LT.	less than
.LE.	less than or equal, i.e., LT|EQ.
.LT|GT.	less than or greater than
.LT|EQ|GT.	less than or equal to or greater than

3.2 Energy Process Hierarchies

In the above, EP's were described as complex events whose constituents are sub-EP's and primitive events (EEV's). It is also possible to define hierarchies of EP's, corresponding to complex events whose constituents are sub-hierarchies and EP's (the latter as primitive events), in the same way as was done above for PP's.

The simplest kind of such a hierarchy is a tree (an event of type EPT), consisting of sub-tree's (SUB-EPT's) and EP's. The tree and its SUB-TREE's each start at a unique EP (the tree root) and end at EP's that are just leaf nodes, not root's of SUB-TREE's. EPT's can be constructed by concatenating the root-EP of the EPT onto a set containing the constituent i-sub-EPT's of the EPT and eventual iSUB-EP's of the root-EP which are not root-EP's.

As specific instances of tree structures, EPT's are represented as EPTx(h)(t) where "h" is the hierarchical index and "t" is the time index of the whole EPTx(h) structure, just as for PPT's above. Thus EPx(k)(t) with k<h designates a SUB-EP of EPx(h)(t) corresponding to the root-EP of a certain SUB-EPT(k)(t) of tree EPT(h)(t). The preceding ideas lead to the following definition of EPT's.

Definition 5. EP's in the context of an EPT

with

a) EPx(h)(t) := an energy process formed through the interaction of its SUB-EP's; it is the root-EP of tree EPTx(h)(t)

b) EPx(j)(t) := a SUB-EP of EPx(h)(t) in the context of tree EPTx(h)(t), formed from interactions among the SUB-EP's in the tree EPT(j)(t) of which EPx(j)(t) is the root-EP)

then

$$EPTx(h)(t)=<\{EPTx(j)(t)|EPx(j)(t),\forall x(j);EPx(j)(t)= EPx(h)(t):iSUB\},EPx(h)(t)>$$

which evidences the 1-to-1 relationship between EPx(h)(t) and EPTx(h)(t), i.e.

EPx(h)(t) .is isomorphic with. EPTx(h)(t)

3.3 Ordered energy value definitions

In refining the above basic energy definitions it is convenient to differentiate between processes that are open and those that are closed with respect to energy flow.

Open systems generally have input and output processes. In thermodynamic systems an input process can correspond to heat flow into the system or work being done on the system and outputs correspond to energy flow out of the system, as heat flow or as work being done by the system.

In thermodynamics the general hierarchical aspect of systems is not emphasized, but it cannot be ignored totally. Indeed, because of energy conservation, input and output energy flows have to come from and go to somewhere. To account for this, the idea of system environment is introduced: the environment is everything in the world that is not part of the particular system under study. In the terminology used above, the system being studied interacts with its environment, the result being the entire world as their SUPER-system.

Closed systems have no inputs or outputs involving energy flow. In an absolute sense only the whole world is a closed system, but the idea of being closed or open relative to specific energy flows within a hierarchy of sub-systems can be useful. Thus classification of a process becomes contextual: $EPx(j)(t)$ can be a process that is open in that it participates in energy flows to or from other EP's in the context of its SUPER-EP, whereas it is closed with respect to energy flows among its own SUB-EP's.

The presence of energy flow for an open EP constitutes evidence for the existence of a "sense" property for such processes, whereas, for a closed EP, no such evidence exists. Sense properties are generally two valued; for an open EP the values would correspond to "input (of energy)" and "output (of energy)". Together the two sense values define a direction in space (Delaney, 2004, Chapt. 5, Sect's. 2.3 and 3.3.1).

Obviously, lack of energy input or output for closed EP's also implies conservation of energy for such processes. Indeed, according to the ideas here presented the principle of energy conservation can be stated in the form

any open EP is a SUB-EP of some closed EP

The preceding ideas lead to the following energy definition in terms of closed root-EP's of EPT's.

Definition 6. F\ord\closed\energy

with

a) EPx(h)(t) := a closed EP, formed through an energy exchanging interaction of SUB-EP's in the context of tree EPTx(h)(t), of which EPx(h)(t) is the root

b) F\ord\closed\energy(EPx(h)(t)) := a function associating a F\ord\closed\energy:value with EPx(h)(t)

c) EPx(h)(t):SUB:Penergy:V := the value of the energy of a SUB-EP of EPx(h)(t), which is assumed non-nil

d) operators .GT. and .EQ. from Table IX

then

1.　　　　　EPx(h)(t):F\ord\closed\energy:value

=

F\ord\closed\energy(EPx(h)(t))

=

EPx(h)(t):Penergy:V
such that

2.　EPx(h)(t):Penergy:V .GT. EPx(h)(t):SUB:Penergy:V

3.　EPx(h)(t):Penergy:V .EQ. EPx(h)(t-1):Penergy:V

Item (3) states that the energy of a closed EP is conserved with variation of time (the index "t").

In item (2), use of the .GT. operator is motivated by mathematical models traditionally used to express the energy of a system as a sum of the energies of its SUB-systems; it assumes that EP's consist of SUB-EP's with non-nil energy values (otherwise .GE. would be required instead of .GT.). The item implies the monotonic increase in energy value with increase of the hierarchical index h in EPx(h)(t); it is analogous to the expression (1) in Section 2.4 that implies monotonic increase in duration value with increase of the time index t in PP's . The energy definition for open EP's is as follows.

Definition 7. F\ord\open\energy

with

a) EPx(h)(t) := a closed EP, formed through an energy exchanging interaction of its SUB-EP's

b) EPx(j)(t) := an open SUB-EP of EP(h)(t), i.e. a SUB-EP of EPx(h)(t) that is open to energy flow in the context of EPx(h)(t), and closed to the flow of energy among its own SUB-EP's (the EP's in the tree EPT(j)(t) of which it is the root-EP)

c) F\ord\open\energy(EPx(j)(t)) := a function associating a F\ord\open\energy:value with EPx(j)(t)

d) EPx(j)(t):SUB:Penergy:V := the value of the energy of a SUB-EP of EPx(j)(t), which is assumed non-nil

then

1.   EPx(j)(t):F\ord\open\energy:value

=

F\ord\open\energy(EPx(j)(t))

=

EPx(j)(t):Penergy:V
such that

2.   EPx(j)(t):Penergy:V .GT. EPx(j)(t):SUB:Penergy:V
and

3.   EPx(h)(t):Penergy:V .EQ. EPx(h)(t-1):Penergy:V
4.   and EPx(j)(t):Penergy:V .LT|EQ|GT.
EPx(j)(t-1):Penergy:V

The preceding definition relative to open EP's is the same as that for closed ones, except for the addition of Item 4. Item 3, the characteristic feature of closed EP's, is included in order to incorporate the principle of energy conservation also into the definition of open EP's. Item 4 contains .LT|EQ|GT. instead of .LT|GT. in order to allow for the case in which energy inputs to and outputs from the EPx(j)(t-1) process cancel each other out, so that there is no net change in the internal energy of the process.

Energy Change Events The above Definitions 6 and 7 can be refined so as to evidence the distinction between open and closed EP's in the structure of their energy values. This involves the introduction of a new kind of primitive event called an energy change event, which permits energy values to be represented as an energy initialization followed by a succession of energy changes. The energy change events are

different for open and closed energy processes: for an open
EP, the value of an energy change event is a pair consisting of
the magnitude of the energy change and its sense (increasing
or decreasing), whereas for a closed EP the value of an energy
change event is just the magnitude of the change. Details are
provided in the following Definition 8.

Definition 8. F\ord\eCh\energy
with
    a) the content of Definition 7
    b) ECH := event type for energy change events
    c) EI :=a special kind of ECH, involving energy
       initialization
    d) recursive definition of EPx(j)(t) as the root of
          EPTx(j)(t) in the context of EPTx(h)(t)
          EPx(j)(t).Penergy.Veval(j)(t)

=

    <EPx(j)(t-1).Penergy.Veval(j)(t-1)|EIx(j)(0).Penergy.
        Veval(j)(0),
      ECH(j)(t).Penergy-change.VdEval(j)(t)>
    e)   for closed EP's
      dEval = energy-change-magnitude:value
    f)   for open EP's
    dEval = (energy-change-sense:value, energy-change-
        magnitude:value)
then
    1.   EPx(j)(t):F\ord\eCh\energy

=

    F\ord\eCh\energy(EPx(j)(t))

=

        EPx(j)(t):Penergy:V

=

    eval(j)(0)|<eval(j)(t-1),dEval(j)(t)>
        such that
   [2.   for closed EP's
   the constraints in items 2), and 3) in Definition 7 are
      respected
    3.   and for open EP's

]   the constraint in item 4) in Definition 7 is also
respected

The conventions for energy change values in Items e) and f) could be duplicated in mathematical models of thermodynamic systems by prefixing a "+" or "-" sign to energy inputs and outputs associated with open systems, such signs being codes for the energy sense value. Closed systems would not require signed energy values.

4. Entropy definition

Traditionally the entropy concept is studied on different levels: on a macroscopic scale in thermodynamics, where it is a variable characteristic of a system's state, and on a microscopic scale in Statistical Mechanics, where it, along with other macroscopic variables, are related to microscopic phenomena, usually in terms of sums or averages of such phenomena. Here only the thermodynamic level is considered.

As in the above study of the energy property, it is useful to distinguish between open and closed systems (processes), in terms of the types of energy flow pertinent to the two process types. For a closed process energy flows are only internal ones and they have the effect (for actual processes) of increasing entropy. Such flows include the case in which the interaction between two processes A and B corresponds to a "perturbation" whereby A supplies a small amount of energy to B, causing the latter to have an inordinately large change of state—this can be viewed as corresponding to a change in B from an unstable equilibrium state to a more stable such state. In the mathematical analysis of such phenomena, process A and the exact magnitude of the perturbing energy is often left out of the model, as if B were a closed process. It is always understood that the relevant closed process is the SUPER process containing A and B.

For open processes, energy flows into and out of the process are also relevant and can cause the entropy of such processes to individually increase or decrease. Thus it is possible to associate an "entropy-sense" property with open processes according to whether their entropy increases or decreases (an analogue of the sense property for energy flow), whereas closed processes have no such sense property—their entropy value just increases.

4.1 Basic entropy definitions

A process having the entropy property will be called an "entropy process". Such processes are distinguished by the type symbol ENTP. For greater generality, they are not defined a-priori to be special cases of the above mentioned processes (ENTP's and PP's).

Formally, ENTP's are complex events whose constituents are sub-ENTP's and primitive entropy change events of type ENTCH. A scenario explaining the exact nature of the ENTCH and ENTP events is as follows: at event ENTEVx(t) the process ENTPx attains a certain equilibrium state with a specific entropy value; it persists in that state and then enters a non equilibrium condition followed by a transition into a new equilibrium state, said transition ending with a new entropy value at event ENTCHx(t+1). Entry into the non equilibrium condition is understood to be the result of an energy input whose value exceeds a certain finite limit such that, although the equilibrium states are generally unstable, they are not exceedingly so. Also the duration of the equilibrium state is understood to be of finite duration.

An ENTP is formed by concatenating an ENTCH onto the end of its immediate sub-ENTP, which leads to the following recursive definition of such processes.

Definition 9. Entropy Processes

ENTPx(t) = <ENTPx(t-1)|ENTCHq(0),ENTCHq(t)>

Basic entropy definitions in terms of ENTP's are presented in the following Table X.

Table X. Basic definitions of entropy in terms of ENTP's

Expression	Meaning
with	
{ENTP.Pentropy.Vent}:= the set containing all ENTP's with entropy value "ent" (the definition set for that value)	
{ENTP.Pentropy.Vent}:inv := the invariant associated with the definition set for value "ent" of the entropy property of an ENTP	
then	
Definition 10. General Entropy	
entropy	{ENTP.Pentropy}:inv, i.e., the characteristic invariant of the set whose elements are ENTP's having some entropy value
entropy:value	{ENTP.Pentropy. Ventropy:value}:inv, i.e., an entropy value "entropy:value"
Definition 11. Entropy as a function	
F\entropy	a function associating a value "F\entropy:value", with an ENTP, i.e., ∀ENTPx.Pentropy, ENTPx:F\entropy:value= F\entropy(ENTPx)
F\entropy:value	the characteristic invariant of the set of ENTP's with which F\entropy associates the specific entropy value "F\entropy:value", {ENTPx; ENTPx:F\entropy:value= F\entropy(ENTPx)}:inv

The general definition defines entropy in terms of the characteristic invariants of sets of ENTP's having the entropy property. As a function, entropy maps an ENTP into a unique entropy value.

4.2 Entropy value ordering

More detailed entropy definitions can be obtained by refining the idea of F\entropy so as to consider specific kinds of such functions. An interesting example would be a function that would evidence the principle of entropy increase in closed ENTP's. The definition of such a function relies on being able to compare entropy values. For such purposes the general method for defining and ordering values described above in Section 2.4 will be used. Thus if some elements of the entropy value definition set {ENTP.Pentropy.Vv} contain elements of the definition set {ENTP.Pentropy.Vv'} and no elements of the latter contain elements of the former then the entropy values are related by the .GT. operator:

{ENTP.Pentropy.Vv}:inv .GT.{ENTP.Pentropy.Vv'}:inv

4.3 Entropy Process Hierarchies

The definition of entropy in terms of ordered values will depend on considering entropy to be a property of ENTP's in the context of entropy process trees (ENTT's), which are structured just like the energy trees mentioned above, as shown in their following formal definition.

Definition 12. ENTP's in the context of an ENTT

with

a) ENTPx(h)(t) := an entropy process formed through the interaction of its SUB-ENTP's; it is the root-ENTP of tree ENTTx(h)(t)

b) ENTPx(j)(t) := a SUB-ENTP of ENTPx(h)(t) in the context of tree ENTTx(h)(t), formed from interactions among the SUB-ENTP's in the tree ENTT(j)(t) of which ENTPx(j)(t) is the root-ENTP)

then

ENTTx(h)(t)=<{ENTTx(j)(t)|ENTPx(j)(t),∀x(j);ENTPx(j)(t)
=ENTPx(h)(t):iSUB},ENTPx(h)(t)>

which evidences the 1-to-1 relationship between ENTPx(h)(t) and ENTTx(h)(t), i.e.

ENTPx(h)(t) .is isomorphic with. ENTTx(h)(t)

4.4 Ordered Entropy Value Definitions

This section presents definitions of entropy in terms of ordered entropy values associated with ENTT's.

First the entropy of closed ENTP's is defined in terms of root-ENTP's of ENTT's.

Definition 13. F\ord\closed\entropy

with

a) ENTPx(h)(t) := a closed ENTP, formed through an energy exchanging interaction of SUB-ENTP's in the context of tree ENTTx(h)(t), of which ENTPx(h)(t) is the root

b) F\ord\closed\entropy(ENTPx(h)(t)) := a function associating a F\ord\entropy:value with ENTPx(h)(t)

c) ENTPx(h)(t):SUB:Pentropy:V := the value of the entropy of a SUB-ENTP of ENTPx(h)(t)

then

1.   ENTPx(h)(t):F\ord\closed\entropy:value

=

F\ord\closed\entropy(ENTPx(h)(t))

=

ENTPx(h)(t):Pentropy:V
such that

2.   ENTPx(h)(t):Pentropy:V .GT. ENTPx(h)(t): SUB:Pentropy:V

3.   ENTPx(h)(t):Pentropy:V .GT. ENTPx(h)(t-1): Pentropy:V

Item (3) is a statement of the Second Law of Thermodynamics for irreversible closed processes. Use of .GT. and not .GE. implies that the change in equilibrium state, evidently towards a more stable state, is accompanied by an increase in entropy. Some situations are excluded from consideration, specifically: processes formed only due to interactions not involving energy flow and so-called "reversible" processes inasmuch as the latter are understood to be idealizations, not actual processes (allowing such processes would require using GE rather than GT).

In Item (2) use of the .GT. operator is motivated by the additivity property of entropy in thermodynamics, which implies the monotonic increase in entropy value with increase of the hierarchical index 'h' in ENTPx(h)(t); it is analogous to the expression (1) in Section 2.4 that implies monotonic increase in duration value with increase of the time index t in PP's .

The entropy of open ENTP's, in the context of ENTT's, is defined as follows.

Definition 14. F\ord\open\entropy

with

    a) ENTPx(h)(t) := a closed ENTP, formed through an energy exchanging interaction of SUB-ENTP's in the context of tree ENTTx(h)(t), of which ENTPx(h)(t) is the root

    b) ENTPx(j)(t) := an open SUB-ENTP of ENTP(h)(t), i.e. a SUB-ENTP of ENTPx(h)(t) that is open to energy flow in the context of ENTTx(h)(t), and closed to the flow of energy among its own SUB-ENTP's (the ENTP's in the tree ENTT(j)(t) of which it is the root-ENTP)

    c) F\ord\open\entropy(ENTPx(j)(t)) := a function associating a F\ord\entropy:value with ENTPx(j)(t)

    d) ENTPx(j)(t):SUB:Pentropy:V := the value of the entropy of a SUB-ENTP of ENTPx(j)(t)

then

    1.   ENTPx(j)(t):F\ord\open\entropy:value

=

F\ord\open\entropy(ENTPx(j)(t))

=

ENTPx(j)(t):Pentropy:V
such that

    2.   ENTPx(j)(t):Pentropy:V .GT. ENTPx(j)(t):
                  SUB:Pentropy:V
                  and

    3. [ENTPx(h)(t):Pentropy:V .GE. ENTPx(h)(t-1):
                  Pentropy:V

    4.]   ENTPx(j)(t):Pentropy:V .LT|GT.
                  ENTPx(j)(t-1):Pentropy:V

Items (2) and (3) are the same as in the definition for closed processes, and are motivated as previously discussed. Item (3) is included to evidence the time (t) dependence of the closed process ENTP(h)(t) in possible contrast with that of its SUB-ENTP's.

Item (4) allows for both increasing and decreasing entropies in transitions to different equilibrium states in open processes, evidently depending on the sense(s) of energy flows into or out of a process. Using .LT|GT. instead of .LT|EQ|GT. follows from the role of t as the index of successive entropy change events.

Detailed Entropy Change Events A useful refinement of Definition 14 is one that more precisely evidences the distinction between open and closed processes, at the level of their constituent entropy change events. In the context of an open ENTP, the value of an entropy change event is a pair consisting of the magnitude of the entropy change and its sense (increasing or decreasing), whereas in the context of a closed ENTP, the value of an entropy change event is just the magnitude of the change. Also, a special kind of entropy change event corresponding to initialization is introduced, permitting entropy values to be represented as an entropy initialization followed by a succession of entropy changes. Details are provided in the following Definition 15.

Definition 15. F\ord\entCh\entropy

with
a) the content of Definition 12
b) ENTI :=a special kind of ENTCH, involving entropy value initialization
c) recursive definition of ENTP of ENTPx(j)(t) as the root of ENTTx(j)(t) in the context of ENTTx(h)(t) (and of entval as its entropy value)

ENTPz(j)(t).Pentropy.Ventval(j)(t)
=
<ENTPx(j)(t-1).Pentropy.Ventval(j)(t-1)|ENTIx(j)(0).
Pentropy.Ventval(j)(0),
ENTCHx(j)(t).Pentropy-change.VdEntval(j)(t)>

d)   for closed ENTP's, dEntval = entropy-change-magnitude:value

e)   for open ENTP's, dEntval = (entropy-change-sense:value, entropy-change-magnitude:value)

then

1.   ENTPx(j)(t):F\ord\entCh\entropy:value

$$=$$

F\ord\entCh\entropy(ENTPx(j)(t))

$$=$$

ENTPx(j)(t):Pentropy:V

$$=$$

entval(j)(0)|<entval(j)(t-1),dEntval(j)(t)>

such that

[2.   for closed ENTP

the constraints in items 2), and 3) in Definition 14 are respected

3.   and for open ENTP's

]   the constraint in item 4) in Definition 14 is also respected

In analogy with the above discussion of energy, closer correspondence with mathematical models could be achieved by adopting the convention that mathematical entropy change values for open processes are prefixed with a "+" or "-" sign according to whether they are increasing or decreasing values, values for closed processes having no such sign.

5. Relations between duration, energy, and entropy processes

The ordering relations between successive values of energies and of entropies mentioned above, together with similar relations for duration (Delaney, 2004, Chapt. 3, Sect. 8.1) are presented together in the following Table XI.

Table XI. Comparison of property value changes in the context of process hierarchies for closed and open processes. Cells contain the operator relating consecutive values with variation of the indices defining process location within a process hierarchy. Entries for energy and entropy assume that the relevant underlying processes arise from interactions involving energy flow.

with
a) V(h)(t):=the value of a certain property in correspondence with node (h)(t) in a process hierarchy
b) V(h)(t).vs.V(h-1)(t) := the order relation between V(h)(t) and V(h-1)(t), e.g., if V(h)(t) is greater than V(h-1)(t) then the order relation is .GT.
c) V(h)(t).vs.V(h)(t-1) := the order relation between V(h)(t) and V(h)(t-1)

then			
property	Duration	Energy	Entropy
Closed process V(h)(t).vs.V(h-1)(t)	.EQ.	.GT.	.GT.
Closed process V(h)(t).vs.V(h)(t-1)	.GT.	.EQ.	.GT.
Open process V(h)(t).vs.V(h-1)(t)	.EQ.	.GT.	.GT.
Open process V(h)(t).vs.V(h)(t-1)	.GT.	.LT|EQ|GT.	.LT|GT.

The table clearly evidences important property similarities. Of particular significance is the above mentioned similarity between: a) the co-variation of energy and entropy values with the h indices of their underlying process trees and b) the co-variation of duration values with the t index of the events in its underlying duration process hierarchy. Also the entropy and duration values vary in the same way with increasing values of the t index in their respective underlying process trees, which is often mentioned as indicating a profound relationship between the two properties. In the present approach the relationship is indirect; it arises from their having the same kind of relationship to their underlying event trees.

The similarity of entropy and energy value variation with the hierarchical index "h" suggests that the formal distinction between energy (EP) and entropy (ENTP) processes may mask a significant overlap between the two structures. Also, the differences in the variation of energy and entropy values with variation of the t index can be largely accounted for by the different ways in which that index is defined for the two properties: for entropy processes it indexes only events involving changes in entropy value and equilibrium states, whereas for energy processes no such restrictions are imposed. Such considerations suggest the hypothesis that ENTP's are just EP's with the entropy property, which is formalized in the following Hypothesis 1.

Hypothesis 1. Entropy processes as special energy processes
    ENTPy(h)(t).Pentropy := EPx(h)(t).Penergy.Pentropy

Acceptance of this hypothesis automatically incorporates energy conservation (the First Law of Thermodynamics) into entropy definitions.

Another interesting hypothesis regards the equivalence of the EP(h)(t)'s and PP(h)(t)'s:

Hypothesis 2. Energy processes as special duration (PP)
    processes
    EPy(h)(t).Penergy:= PPx(h)(t).Pduration.Penergy

It would seem possible to accept Hypothesis 2 as a relationship between processes (primary existents) if EP's do always have the duration property, which seems eminently reasonable since otherwise the idea of energy conservation would lose all significance.

Evidently, acceptance of both Hypotheses 1 and 2, would imply that ENTP's are special PP's. This would explain the co-variation of duration and (closed process) entropy values with time as being due to their genesis in the same kind of event structure.

It is also interesting to speculate on the possibility that EP(h)(t)'s and PP(h)(t)'s are equivalent, i.e.:

Hypothesis 3. Equivalence of energy and duration processes
EPy(h)(t).Pduration.Penergy = PPx(h)(t).Pduration.Penergy

Rejection of Hypothesis 3 after acceptance of Hypothesis 2 would require the demonstration of the existence of a PP(h)(t) whose hierarchical structure does not rely entirely on interactions involving energy flow.

6. Summary

After illustrating notational conventions for defining and describing primary and secondary existents and their application to the definition of duration as a property of a specific "duration process" type of primary existent, particular attention was devoted to other kinds of primary existents pertinent to the definition of the properties of energy and entropy. For both properties, definitions were developed at various levels of detail, in terms of existents corresponding to simple processes, formally distinct from duration processes, and to hierarchies of such processes.

Energy definitions were formulated in terms of energy processes in such a way as to incorporate the principle of energy conservation.

Entropy definitions in terms of entropy processes were formulated so as to incorporate constraints imposed by the Second Law of Thermodynamics. Ongoing entropy increase in closed systems emerged naturally because of how entropy values were related to entropy process structure in the definitions, which also implied the absence of an entropy change sense for closed systems.

Comparisons of energy processes and entropy processes suggested that entropy processes are just energy processes that also have the property of entropy, which automatically incorporate energy conservation (the First Law of Thermodynamics) into entropy definitions.

The comprehensive nature of the proposed energy and entropy definitions and their agreement with perceptual and mathematical models considered to be well validated by empirical data, confirms the generality and power of the methodology followed in their construction.

The possibility that energy processes are special cases of, or are even equivalent to, duration processes was also investigated and formulated in terms of hypotheses whose acceptance could offer a plausible explanation for correlations in observed entropy and time value variations.

9

LOCALITY, RELATIVITY, AND
THE ERP PARADOX

1. Introduction

The ERP experiment (Einstein A, Podolsky B, Rosen N, 1935) is often characterized as implying the possibility of non local phenomena in Physics leading to an incompatibility of Quantum Mechanics and Special Relativity. The experiment involves N repetitions of the following process: a particle $Q(0)$ with zero spin angular momentum decays into two fermions, $Q(1)$ and $Q(2)$, having equal mass and traveling equal distances in opposite senses towards spin measurement analyzers: A_1 for $Q(1)$ and A_2 for $Q(2)$. Symbols $s(1)$ and $s(2)$ denote the spin projection values along a specific axis measured by the analyzers respectively for $Q(1)$ and $Q(2)$. $MP(1)$ and $MP(2)$ denote the measurement procedures by which $s(1)$ and $s(2)$ are measured.,The magnitude of the spin projection values is $s= \frac{1}{2}$ in the system of units where Plank's constant $h=2\pi$.

Each repetition of the just described process yields $s(1) = -s(2)$ as would be expected for angular momentum conservation. Also, for a large number of repetitions, $s(1) = +1/2$ about half the time and $-1/2$ the other half, the same thus being true for $s(2)$, i.e. the two values $\pm \frac{1}{2}$ have equal probability for both particles. The type of experiment under consideration will be referred to below as an ERP experiment.

The experimental results give rise to a dilemma: if the result for each particle is governed only by probabilistic considerations as anticipated in Quantum Mechanics, how does it turn out that s(1) = - s(2) in each step of the experiment—how do Q(1) and Q(2) 'know' each other's result so as to be able to always obtain the opposite spin projection value ? It would seem as if there might exist an 'instantaneous' interaction between the spin analyzers. If their separation, D, is very large, this could require faster-than-light communication—in the limit instantaneous communication, which is synonymous with non-locality.

The analysis of the experimental results is based on two principles:

Principle 1. from Quantum Mechanics, a particle does not 'have' a spin projection value until it is measured

Principle 2. from Special Relativity, the velocity of a signal does not exceed that of light

One possible implication of the above Principle 2 is to consider that the ERP experiment to be a test of the following hypothesis

Hypothesis 1: Space time locality of interactions

For large separation of analyzers A(1) and A(2), the measurement at one of the analyzers for one of the products of the decay of Q(0) cannot "influence" the measurement at the other analyzer for the other product of the same decay.

Hypothesis 1 test strategy: After setting up the experiment such that the distance D between the analyzers would require faster than light communication between the spin measurement events at the analyzers in order for them to share information about their outputs for the same decay event, perform a measurement that can determine if the s(1) and s(2) values obtained in the repetitions of the experiment are random and satisfy s(1) = - s(2) in all cases.

The results of the test strategy are positive: in each step of the experiment, s(1) = - s(2) irregardless of the magnitude of D, which is often interpreted as requiring the existence of a faster than light signal connecting the analyzers—although no such signal is directly observed—and thus as evidence suggesting the falsification of locality hypotheses such as the one defined above. This leads to the conclusion that such experiments evidence an incompatibility of Quantum Mechanics and relativity since traditional Physics lore would have it that the latter requires locality whereas the former does not. Such conclusions can however be avoided using a relativistic formulation in the context of the Discrete Event Physics (Delaney, 2005) paradigm. The following Section 2 explains this approach in detail and Section 3 summarizes the results.

2. Non local relations in the ERP thought experiment

Einstein's theories of relativity are theories about relations between values of properties and reference frames which are events or systems to which the values are referred. Such relations can be formalized as <fr, value> where 'value' is the value of a property in the frame of reference 'fr'. A frame sensitive property of some entity (system or event), becomes multi-valued if its frame is not specified, i.e. if the relation as a whole is not specified—which is an argument in favor of the point of view that a (frame sensitive) property is best understood to be 'of' a relation and not 'of' one of its related entities.

The above description of the ERP experiment can be reformulated in terms of relations and their properties as follows.

1. spin = the spin property, as projected on a certain direction
2. q(0) = the decay event of particle Q(0)
3. q(1) = <q(0), MP(1)>.Pspin.Vs(1) = a relation q(1) between system q(0) and the measurement procedure MP(1);

relation $q(1)$ is the motion of fermion $Q(1)$ and has property 'spin' with value $s(1)$ as measured by $MP(1)$. The possible values of $s(1)$ are $s(1) = s_1 = \frac{1}{2}$ and $s(2) = s_2 = -\frac{1}{2}$,,

4. $q(2) = <q(0), MP(2)>.Pspin.Vs(2) = $ a relation $q(2)$ between system $q(0)$ and the measurement procedure $MP(2)$; relation $q(2)$ is the motion of fermion $Q(2)$ and has property spin with value $s(2)$ as measured by $MP(2)$. The possible values of $s(2)$ are $s(2) = s_1 = \frac{1}{2}$ and $s(2) = s_2 = -\frac{1}{2}$

5. $q(1,2) = <q(1),q(2)>.Pspin.Vs(1,2) = $ a relation $q(1,2)$ between relations $q(1)$ and $q(2)$ such that $s(1)=inverse(s(2))$ so $s(1,2)=0$—this relation being an expression of angular momentum conservation

A mathematical model of the probability property of relation $q(1,2)$ is the joint probability distribution in Table I below, from which the marginal distributions in that table can be derived for relations $q(1)$ and $q(2)$, using standard methods of probability theory: for $i \in [1,2]$ the marginal probability for the i-th value of s for the j-th relation $q(j)$ is

$$Pr(s_i (j)) = Pr(s_i (j), s_1(k)) + Pr(s_i (j), s_2(k)),$$

where $Pr(s_i (j), s_n(k))$ is the joint probability of $s_i (j)$, and $s_n (k)$.

Table I. The joint probability $P(s(2), s(1))$,of the values $s(1)$ and $s(2)$ of the spin property of two fermions and the conditional $Pr(s(2)|s(1)_,$ and marginal probabilities $Pr(s(i))$, $i \in [1,2]$ that can be derived from the joint probabilities

,	Marginal Pr(s(1)),	Marginal, Pr(s(2))		Conditional Pr(s(2)|s(1))_,		Joint Pr(s(2),s(1))	
s(2) ＼ s(1)		$\frac{1}{2}$	$-\frac{1}{2}$	$\frac{1}{2}$	$-\frac{1}{2}$	$\frac{1}{2}$	$-\frac{1}{2}$
$\frac{1}{2}$	.5			0.	1.	0.	.5
$-\frac{1}{2}$	.5			1.	0.	.5	0.
		.5	.5				

Such a model is consistent with the above described results of the ERP experiment. The results do not furnish evidence for rejection of Hypothesis 1. Whether they furnish evidence for its acceptance or not is questionable. This depends on the exact meaning of the word 'influence' as used in that hypothesis: if it is to be understood as to imply faster than light energy transfer, the hypothesis in question would seem to be irrelevant to the ERP experiment in the Discrete Event Physics paradigm.

The results of ERP-type experiments do provide evidence for non locality because the relations involved are non local in that they are composed of events separated in space time. This is analogous to relations in Einstein's theories of relativity that involve reference to measurements of events referred to distant reference frames, examples being measurements of spatial coordinates, energy, momentum, etc . The basic point is that relativity foresees non locality for some *relations* but not for *interactions*.

On the basis of the preceding it seems that Hypothesis 1 is not really appropriate for the ERP experiment. Recalling that relations are a special case of SUPER events in Discrete Event Physics, the following hypothesis is suggested as an alternative to Hypothesis 1.

Hypothesis 2
There does not actually exist a multi layer hierarchy of events, having relations as a special case

The expected results of the ERP experiment would seem to furnish evidence for the rejection of this latter hypothesis in that the experiment involves three event levels:

- level 1 containing the systems Q(0), A(1) and A(2)
- level 2 containing the SUPER relations q(1) and q(2)
- level 3 containing the SUPER relation q (1,2)

The mathematical model presented in Table I does not evidence the existence of signals that transmit energy or information between systems Q(1) and Q(2), essentially because the model does not refer directly to those systems but to relations q(1), q(2) and q(1,2). In order to try to further clarify the nature of the relations q(1) and q(2) one can model them as Quantum Mechanical waves eminating from the decay event Q(0). Such waves can be either plane waves or spherical waves. Using plane waves suggests the apparent need for non-local communication between Q(1) and Q(2) as they separate from Q(0). Such communication is not necessary in the case of q(1) and q(2) as spherical waves because the latter would be superposed at all times.

3. Summary

After introducing the ERP paradox with reference to an experiment that seems to lead to incompatibility of specific fundamental principles of Quantum Mechanics and Special Relativity and to imply non locality of interactions, an alternative approach to the analysis of such experiments was presented that conserves both principles. The approach emphasizes the importance of recognizing the existence of space-time extended relations between events—analogous to the relations in Special Relativity connecting events to reference frames.

Specifically, it was argued that, just as changing the reference frame to which the velocity of an object is referred in Special Relativity will result in a new velocity value for the object, a change in one of the SUB events of a relation can result in a new relation with a new value of the same relation had by both the new and old relations. This effect can explain the results of the ERP experiment without violation of the above mentioned principles, so that no paradox arises. Depending on how it is defined, non locality (of relations) remains a possibility, but this is a typical characteristic of relativity that does not necessarily imply 'action at a distance'.

10

LOCALITY IN DOUBLE SLIT INTERFERENCE
EXPERIMENTS

Arguments are presented in support of the argument
that non locality of interactions is not a necessary part of
interpretations of double slit interference experiments, at least
not for what regards the spatial distribution and brightness of
resulting illuminated areas.

1. Introduction

Results of experiments involving interference gratings that
supported the theory of 'wave-like behavior' were fundamental
in the development of physical optics, where, for example
Young's interference experiment led to the acceptance
of the wave interpretation of light as compared to the
previously favored corpuscular interpretation due largely to
Newton. Similar interference experiments involving massive
particles had a similar effect on the development of quantum
mechanics.

Such experiments seem however to present a problem that
is largely ignored in the literature. They can be interpreted
as support for arguments in favor of non local interactions:
spatial non locality, temporal non locality or both.

It is interesting how the non locality problem has been
ignored for what regards interference experiments involving
gratings while it has received widespread attention in
discussions of tests of Bell's inequalities (Bell, 1966). Indeed

the latter such discussions often assert that the tests provide very strong arguments in favor of the necessity of non locality in any fundamental theory of Physics.

The objective of this article is to demonstrate that the results of experiments involving interference gratings do not support arguments in favor of the necessity of non local interactions for their explanation.

2. The double slit experiment

This section discusses the setup and results of the double slit experiment involving light and the various interpretations of the results with regard to their implications regarding non locality.

2.1 Setup and results

The general setup of an interference experiment is visualized in Figure 1 for the special case of a grating containing two parallel slits. A source S repeatedly emits photons that all move towards the grating with the same speed. A photon is idealized as being a wave front at the grating; it gives rise to Huygens wavelets that leave the grating slits and travel toward a screen where its energy is deposited in a small area ("detection point" P). In the following 'wavelet' is also intended to signify wave front.

Figure I. The double slit experiment setup. Wavelets from two slits (Slit 1 and Slit 2) on a grating 'interfere' at a 'detection point' P on a screen. Lines (R,r,ρ) connect points (BP,AP,CP) and represent the distances between their extremities.

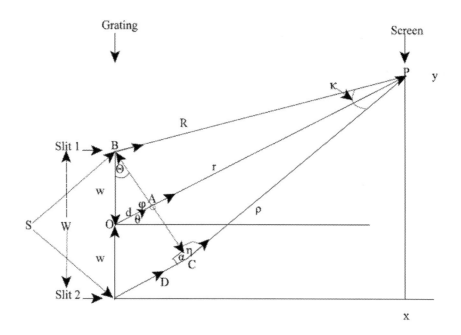

The detection points of consecutive particle-waves correspond to a pattern of oscillating point density on the screen: illumination bands separated by bands of darkness. If one of the slits is covered the illumination becomes more uniform, with no more bands. Uncovering the slit, the band structure reappears.

2.2 Locality problems and possible solutions
The major difficulty in understanding the band structure in the surface illumination is that uncovering the covered slit causes the illumination to decrease at the locus of the dark bands. How can an increase in the transparency of the grid reduce illumination at the locus of a dark band? The standard answer relies on modeling the light leaving the grid as consisting of wavelets, with one wavelet leaving each slit and

expanding radially outward from the slit towards the screen, with the result that if the wavelets leave the slits at the same time and arrive at the same time at some specific point on the screen—point P in Figure 1—, they may tend to combine differently at different such points so as to result in the above mentioned band structure.

The combination of the wavelets involves various factors:

- each wavelet is assumed to have a probability amplitude property, modeled here as a pair of properties (amp-sign, amp-magnitude) with a corresponding pair of property values (amp-sign-v, amp-mag-v), which will be abbreviated in the following to (sign, magnitude) where
- the magnitude value varies periodically as the wavelet expands, changing repeatedly from a maximum to a minimum value and then back to the maximum value (like a cosine function does between 0 and 2π)
- there are two possible sign values, which are each other's inverses, the sign inverts each time the magnitude attains its minimum value
- the probability that the photon's energy will be deposited at a point on the wavelet is a function of the magnitude value of the wavelet amplitude property at that point
- if two wavelets overlap at some point they generally 'collapse' and give rise to a new wavelet with its own probability amplitude property, whose sign and magnitude are functions of those of the overlapping wavelets

A problem with the preceding ideas is that if the wavelets from the two slits leave the slits at the same time and travel at the same velocity they will, in general, not arrive at point P at the same time, except for the case in which point P is at the center of the band structure on the screen. In the following three sub sections approaches for overcoming this problem will be studied.

2.2.1 wavelets as wave train constituents

It is assumed that sequences of wave fronts, collectively called wave trains, emanate from the spatial location of source S in Figure 1 and from all wavelet emanation points such as Slits 1 and 2 in the figure. Train Ψ_1 passes through Slit 1, and,train Ψ_2 passes through Slit 2. The j-th wave front in Ψ_1 , and the k-th ,wave front in Ψ_2 , then overlap at point P on the screen. If the distance from S to P through Slit 1 is not equal to that from S to P through Slit 2 and if both fronts travel at the same speed, v, and arrive at P at the same time T it is necessary that the events in which they are emitted from the source S occur at different times (but at the same spatial location). The time differences at the source can be evaluated as follows.

Letting

Ψ_1 and Ψ_2 = the wave trains respectively emanating from points passing through slits 1 and 2

λ = the distance between consecutive fronts in both trains

τ = the time interval between consecutive fronts in both trains

$v = \lambda/\tau$ = the propagation velocity of all fronts in both trains

ψ_{1j} = the j-th wave front in Ψ_1

ψ_{2k} = the k-th wave front in Ψ_2

T = the time at which both ψ_{1j} and ψ_{2k} arrive at point P on the screen

$d_1(j) = j\lambda + \delta_1$, for integer j and $0 \le \delta_1 < \lambda$, = the path length from point S, through Slit 1 to point P on the screen,

$t_1(j) = T - d_1(j)/v$ = the time at which front ψ_{1j} in Ψ_1 emanates from point S

$d_2(k) = k\lambda + \delta_2$, for integer k and $0 \le \delta_2 < \lambda$, = the path length from point S, through Slit 2 to point P on the screen

$t_2(k) = T - d_2(k)/v$ = the time at which front ψ_{2k} in Ψ_2 emanates from point S

then

the magnitude of the difference in the emanation times is

$$\delta t = | t_2 (k) - t_1 (j) | = | [d_1 (j) - d_2 (k)]/v | \qquad (0a)$$
$$= | (j-k)\tau + (\delta_1 - \delta_2)/v | \qquad (0b)$$

and the magnitude of the difference in the path lengths from the slits to point P is

$$\delta d = | d_2 (k) - d_1 (j) | = v\delta t = | (j-k)\lambda + (\delta_1 - \delta_2)| \quad (0c)$$

A basic property of a wave train Ψ is its periodicity, i.e. with ψ_n ,being the nth wave front in Ψ (counting from 0), such that, as a function of the front's phase angle 'a',

$$\psi_0 (a) = e^{ia} = \cos(a) + i \sin(a)$$

so that for integer n

$$\psi_{2n} (a) =, \psi_n (a + 2\pi n) = \psi_0(a)$$
$$\psi_n (a) = \psi_0 (a+\pi n) = - \psi_0 (a)$$

Both the real and imaginary parts of ψ have values that oscillate repeatedly between a maximum positive value and a minimum negative value, passing through zero midway between those values. This implies that if the fonts ψ_{1j} in train Ψ_1 and ψ_{2k} in train Ψ_2 overlap at point P in Figure 1 as described above, and therefor combine to form a super wave front, then

$$\psi = \psi_{1j} + \psi_{2k} \qquad (1)$$

The preceding considerations regarding the overlapping of ψ_{1j} and ψ_{2k} suggest one of the following consequences:

- CASE 1: *constructive interference* where ψ_{1j} and ψ_{2k} have the same sign, so that they reinforce each other in the sense that $|\psi| > | \psi_{1j} |$ and $|\psi| > |\psi_{2k}|$—which corresponds to the distances δ_1 and δ_2 in Equation (0b) being either both $\geq \lambda/2$ or both $< \lambda/2$. In the special cases that $\delta_1 =\delta_2 =\lambda/2$ or $\delta_1 =\delta_2 = 0$, then $\delta_1 - \delta_2 = 0$, which is called *maximum constructive* interference—it occurs along the center of an illuminated band on the screen.

- CASE 2: *destructive* interference in which ψ_{1j} and ψ_{2k} have opposite signs, in which case $|\psi|$ is less than the larger of $|\psi_{1j}|$ and $|\psi_{2k}|$. In this case either $\delta_1 > \lambda/2$ and $\delta_2 < \lambda/2$ or vice versa $\delta_1 < \lambda/2$ and $\delta_2 > \lambda/2$. In the special cases that $\delta_1 = -\delta_2 = \lambda/2$ or $\delta_1 = -\delta_2 = -\lambda/2$, then $|\delta_1 - \delta_2| = \lambda$, which is called *maximum destructive interference*—it occurs along the center of a dark band on the screen.

Since Quantum Mechanics postulates that the probability of point P being in an illuminated area on the screen is $|\psi|^2 = \psi'\psi$ where ψ' is the complex conjugate of ψ, constructive interference increases that probability while destructive interference decreases it. Such considerations emphasize the importance of the <u>signs</u> of fronts ψ_{1j} and ψ_{2k} in determining the value of the probability of energy deposition at point P, large and small values respectively implying that P is in a strongly illuminated or in a very dark area on the screen.

The position of the centers of the illuminated bands on the screen can be determined in terms of three elements: 1) CASE 1 and CASE 2 above, 2) Figure 1, and 3) the following mathematical model:

$$\tan(\theta) = y/x,$$
$$\alpha = \pi/2$$
$$\sin(\Theta) = D/W$$
which, in the limit $\Theta \to \theta \to 0$, implies
$$y/x = \tan(\theta) \to \sin(\theta)$$
so that the position y of a band's center on the screen at x is
$$y = x \times D/W$$
and, more generally, for the n-th band, corresponding to
$$D_n = n\lambda \text{ is}$$
$$y_n = x \times D_n /W = x \times n\lambda/W$$

2.2.2 massive virtual photon approach

Assuming, in contrast with the previous sub section, that only single wave fronts (wave trains consisting of only one wave front) emanate from the slits and that the wavelets leave the slits at the same time, the problem becomes how to determine what are the velocity values of the wavelets required for their simultaneous arrival at the energy deposition point on the screen, and how they might come to have such values. Two approaches to solving this problem will be considered, one here and one in the following sub section.

The first approach is similar to derivations traditionally used to predict the location of the illumination bands in the double slit experiment. It predicts the same band positions but also treats the above problem concerning the simultaneity of the wave arrivals at point P. Specifically it suggests the that photons leaving the slit closest to the illuminated point (P in Figure 1) have a larger mass than those leaving the furthermost slit, and therefor travel slower than the later. To justify such a suggestion it can be argued that the photons are not "free" but "virtual" ones, since they are confined to a finite space. The mass of the slower photon is determined such that the two photons arrive simultaneously at P, as follows.

With reference to Figure 1, and using the convention that r, R, ρ are symbols representing both spatial trajectories and their lengths, and defining

$$\alpha = \pi/2 \tag{1}$$

then

$$R^2 = r^2 + (w\cos(\Theta))^2 \tag{2}$$

and also defining

$$\eta = \pi/2 \tag{3}$$

then

$$R^2 = \rho^2 + (W \cos(\Theta))^2$$

and, in the limit R>>W=2w

$$R=\rho=r \qquad\qquad (4)$$

Also, with v<c being the velocity along R, the transit time along that trajectory is

$$T=R/v$$

and with c being the velocity over distances D and ρ, the transit time over said distances is

$$T'=(D+\rho)/c$$

Thus if

$$T'=T$$

then

$$c/v=(\rho+D)/R$$
$$v/c=1/(1+D/R)$$

and with R>>D

$$v/c=1-D/R$$
$$\gamma^2=1/(1-(v/c)^2) = R/(2D)$$

and with the relativistic energy expression

$$E=m\gamma c^2$$

then

$$(E/(mc^2))^2 = R/(2D)$$

Using the expression for the heavy photon velocity v in terms of its de Broglie wavelength λ and frequency f

$$\lambda f=v$$

the quantum mechanical expression(h= Plank's constant) for the energy

$$E=hf$$

yields

$$mc^2 = (2D/R)^{1/2}(hf/\lambda)$$

for the mass associated with the trajectory R, so that

$$(R\rightarrow\infty) \text{ implies } (m\rightarrow 0)$$

screen location of bands At P the wavelets from B and C collapse and give rise to a super system having the property of probability amplitude, whose value depends on the value of the wavelength λ associated above with the wavelet expanding along the trajectory labeled "D" from B' to C as in the following cases:

- CASE 3: the length $D= n\lambda$ for integer n so that the amplitudes originating at Slit 2, and at C have the same sign values, and the amplitudes from Slits 1 and 2 are <u>therefor</u> said to combine *constructively* at P—reinforcing each others effect so as to yield a relatively bright band on the screen—as in the case of maximum constructive interference defined in CASE 1 above
- CASE 4: the length $D=n\lambda/2$ for integer n so that the amplitudes originating at Slit 2, and at C have different sign values, and the amplitudes from Slits 1 and 2 are <u>therefor</u> said to combine *destructively* at P—diminishing each other's effect, so as to yield a relatively dark band on the screen—as in the case of maximum destructive interference defined in CASE 2 above

With reference to the geometry of Figure 1, a mathematical model incorporating the descriptions cited in cases 3 and 4 is as follows:

$$\tan(\theta) = y/x$$

and using (1), i.e.

$$\alpha = \pi/2$$

then

$$\sin(\Theta) = D/W$$

so that, in the limit $\Theta \to \theta \to 0$,

$$\tan(\theta) \to \sin(\theta)$$
$$y/x = D/W$$

Since, for large R and small θ, D is approximately the path difference to P from the slits, then if

$$D = n\lambda \text{ for integer } n$$

then y has a high probability of being in the nth illuminated band from the center of the interference pattern.

Thus, according to the discussions under CASE 3 and CASE 4 above, if

$$D = n\lambda \text{ for integer n}$$

then y has a high probability of being in the nth illuminated band from the center of the interference pattern, whereas if $D=n\lambda/2$ for integer n then y has a high probability of being in the nth non illuminated band from the center of the interference pattern

A criticism of the above approach is that its basic assumption suggests that both photons should be treated as being massive since they are both confined. Of course, the more confined one, moving along R, could be expected to have the larger mass and slower velocity, in support of the idea that they can thus arrive simultaneously at P.

2.2.3 massive super system approach

An alternative to the approach in the immediately preceding section involves considering the two-photon system (one photon from each slit) as a whole (super system) that has a non zero effective mass and moves slower than its constituent photons (sub systems). The super system coincides with the center of inertia system of its constituent photons, in which the momentum vectors of the latter add to zero.

The center of inertia and the individual photons are assumed to follow specific trajectories in Figure 1. The center of inertia super system propagates from point O to point P along the trajectories labeled 'd' and 'r'. One of the photons propagates from point B to P along R. More trajectory details will be discussed after the ensuing discussion of the super system motion, during which the following conventions are adopted: primed symbols represent values of properties in the center of inertia system, and bold case symbols represent values of properties of the super system. Thus, with c = the velocity of light

e(i) = the energy of the ith photon in the laboratory reference system

Q(i) = the momentum 3-vector of the ith photon in the laboratory reference system (that of the grating and screen)

q(i) = the magnitude of Q(i)

μ(i) = the magnitude of the velocity of the ith photon

e = e(1)=e(2)

q = q(1)=q(2)

μ = μ(1)=μ(2)

r, R, ρ are symbols representing both spatial trajectories and their lengths

then

qc = e

μ = qc/e

and with

U= the momentum 3-vector of the 2-photon super system in the laboratory reference system

U = Q(1) + Q(2)

u = the magnitude of U

e = the energy of the 2-photon super system in the laboratory reference system

V= the velocity vector of the center of inertia in the laboratory reference system,

along the trajectory connecting points A and P in Figure 1

v = the magnitude of V

then

e = e(1)+e(2) = 2e

v= uc/e

and with

κ = the angle between the photons in the laboratory system (a property of the two photon system)

then
$$V^2 = (Uc)^2/e^2 = 2(qc)^2/e^2 + 2(qc)^2\cos(\kappa)/e^2$$
$$= 2(qc)^2 (1+\cos(\kappa))/4e^2$$
$$= \mu^2(1+\cos(\kappa))/2$$
$$= \mu^2 \cos^2(\kappa /2)$$
$$(V/c)^2 = \cos^2(\kappa/2)\times(\mu/c)^2$$
$$\gamma^2 = 1/[1-(v/c)^2] = 1/[1-\cos^2(\kappa/2)\times(\mu/c)^2]$$

For $\mu=c$

$$v/c = \cos(\kappa/2)$$
$$\gamma^2 = 1/[1-\cos^2(\kappa/2)] = 1/\sin^2(\kappa/2)$$
$$\gamma = 1/\sin(\kappa/2)$$
$$e = mc^2\gamma = mc^2/ \sin(\kappa/2)$$

From Figure 1
$$\sin(\kappa/2) = \cos(\Theta)u/R$$
so $e = mc^2R/(u\times\cos(\Theta))$
$$= mc^2R/u \text{ as } \Theta \to 0$$
and with
ν=the frequency of the de Broglie wave associated with the super system in the lab
then
$$e = h\,\nu = mc^2R/u$$
$$mc^2 = h\,\nu\,u/R$$
and with
ν' = the frequency of the de Broglie wave associated with the super system in the center of inertia frame (its own rest frame)
then
$$\nu = \nu'\gamma$$
$$mc^2 = h\,\nu'$$

A problem with the immediately preceding presentation is that it is not obvious how it eliminates temporal non locality: if r/R=v/c then the super system arrives at P at the same time that the photon along R arrives there. But a photon propagating along D and ρ will arrive later.

A solution can be achieved by a judicious choice of the photon trajectories. With reference to Figure 1 and the arrows along the various trajectories it depicts, the photons (wavelets having wavelength λ) from the slits (B and C in the figure) expand radially and overlap at the point O on the grating midway between the slits, where their partial superposition at O gives rise to their 'collapse' and to the creation of a super system, corresponding to a wavelet with wavelength Λ expanding radially from O. The wavelet centered at O expands to A along the trajectory labeled "d". At A it collapses and a new wavelet with wavelength λ expands radially from that point towards points B and C—at which points it arrives simultaneously and collapses, giving rise to new wavelets with wavelength λ expanding radially from the latter points. The new wavelets expand to P: from B along R, and from C along ρ, where ρ and R have equal lengths. Obviously the paths followed by the photons from the two slits are of equal length, so the photons (wavelets) will arrive at P at the same time if they travel at the same velocity c. Denoting the super system velocity along r by v, if $r/R=v/c$ the super system and the two photons arrive simultaneously at P.

screen location of bands At P the wavelets from B and C collapse and give rise to a super system having a probability amplitude property whose value depends on the value of the wavelength Λ associated above with the wavelet expanding along the trajectory labeled "d" from O to A as in the following cases:

- CASE 5: the length $d= n\Lambda= n\lambda/2$ for integer n so that the amplitudes associated with the super system at O and at A have the same sign values, and the amplitudes from Slits 1 and 2 are <u>therefor</u> said to combine *constructively* at P—reinforcing each other so as to yield a relatively bright band on the screen—as in the case of maximum constructive interference defined in CASE 1 above
- CASE 6: the length $d=2n\Lambda=n\lambda$ for integer n so that the amplitudes originating at O, and at A have different

sign values, and the amplitudes from Slits 1 and 2 are therefor said to combine destructively at P—diminishing each other's effect, so as to yield a relatively dark band on the screen—as in the case of maximum constructive interference defined in CASE 2 above

With reference to the geometry of Figure 1, a mathematical model incorporating the descriptions cited in cases 5 and 6 is as follows:

$$\tan(\theta) = y/x$$

and using (instead of $\alpha=\pi/2$, as in the previous sub-section)

$$\phi = \pi/2$$

then

$$\sin(\Theta) = d/w$$

so that, in the limit $\Theta \to \theta \to 0$,

$$\tan(\theta) \to \sin(\theta)$$
$$y/x = d/w$$

According to the discussions under CASE 3 and CASE 4 above then, if

$$d = n\lambda/2 \text{ for integer } n$$

then y has a relatively high probability of being in the nth illuminated band from the center of the interference pattern, whereas if

$$d=n\lambda \text{ for integer } n$$

then y has a relatively high probability of being in the nth non illuminated band from the center of the interference pattern.

A possible objection to the proceeding trajectory choices is that it involves rather unusual uses of Hughens wavelets, assumed to emanate not only from apertures but also from other locations in space. However the usual statement of Huygen's Principle does not preclude such phenomena in that it asserts that *all* points on any wave front of light are sources of wavelets.

3. Observations

The above presentation does not evidence that interpretations of the results of interference experiments in ways relying on non local interactions are incorrect in some way. It only evidences that credible local interpretations are possible, so that non local alternatives are not necessary.

REFERENCES

Einstein A, Podolsky B, Rosen N, *Can Quantum Mechanical Description of Reality be considered Complete?*, Physical Review 47 777 (1935)

Delaney W, *Discrete Event Physics*, iUniverse, New York, (2004, 2005)

Bridgeman P.W., *The Logic of Modern Physics*, New York (1927)

Delaney W. (1999) "Limitation of Operational Definitions", Int. J. Theor. Phys. vol 38, p 1757.

J. S. Bell, *On the problem of hidden variables in quantum mechanics*, Rev. Mod. Phys. **38**, 447 (1966)